COMMUNITY DESIGN

Cities & Planning Series

The ***Cities & Planning Series*** is designed to provide essential information and skills to students and practitioners involved in planning and public policy. We hope the series will encourage dialogue among professionals and academics on key urban planning and policy issues. Topics to be explored in the series may include growth management, economic development, housing, budgeting and finance for planners, environmental planning, GIS, small-town planning, community development, and community design.

W. Arthur Mehrhoff

COMMUNITY DESIGN

A Team Approach to Dynamic Community Systems

Cities & Planning

SAGE Publications
International Educational and Professional Publisher
Thousand Oaks London New Delhi

For information:

SAGE Publications, Inc.
2455 Teller Road
Thousand Oaks, California 91320
E-mail: order@sagepub.com

SAGE Publications Ltd.
6 Bonhill Street
London EC2A 4PU
United Kingdom

SAGE Publications India Pvt. Ltd.
M-32 Market
Greater Kailash I
New Delhi 110 048 India

Printed in the United States of America

Library of Congress Cataloging-in-Publication Data

This book is printed on acid-free paper.

Mehrhoff, W. Arthur.
Community design: A team approach to dynamic community systems / by W. Arthur Mehrhoff.
p. cm. — (Cities and planning; v. 4)

Includes bibliographical references and index.
ISBN 0-7619-0596-0 (cloth: acid-free paper)
ISBN 0-7619-0597-9 (pbk.: acid-free paper)
1. City planning—United States. 2. Land use, Urban—United States. 3. Community development—United States. 4. Urban policy—United States. I. Title. II. Series: Cities & planning series; v. 4
HT165.52 .M45 1998
307.1'2'0973—ddc21 98–25434

99 00 01 02 03 04 05 7 6 5 4 3 2 1

Acquiring Editor: Catherine Rossbach
Editorial Assistant: Heidi Van Middlesworth
Production Editor: Diana Axelsen
Editorial Assistant: Stephanie Allen
Typesetter: Christina Hill

CONTENTS

SERIES EDITORS' INTRODUCTION

The study of cities is a dynamic, multifaceted area of inquiry that combines a number of disciplines, perspectives, time periods, and actors. Urbanists alternate between examining one issue through the eyes of a single discipline and looking at the same issue through the lens of a number of disciplines to arrive at a holistic view of cities and urban issues. The books in this series look at cities from a multidisciplinary perspective, affording students and practitioners a better understanding of the multiplicity of issues facing planning and cities and of emerging policies and techniques aimed at addressing those issues. The series focuses on traditional planning topics, such as economic development, management and control of growth, and geographic information systems. It also includes broader treatments of conceptual issues embedded in urban policy and planning theory.

The impetus for the *Cities & Planning* series was our reaction to a common recurring event—the ritual selection of course textbooks. Although we all routinely select textbooks for our classes, many of us are never completely satisfied with the offerings. Our dissatisfaction stems from the fact that most books are written for either an academic or practitioner audience. Moreover, on occasion, it appears as if this gap continues to widen. We wanted to develop a multidisciplinary series of manuscripts that would bridge the gap between academia and professional practice. The books are designed to provide valuable information to students/instructors and to practitioners by going beyond the narrow confines of traditional disciplinary boundaries to offer new insights into the urban field.

Arthur Mehrhoff's *Community Design: A Team Approach to Dynamic Community Systems* represents a unique way of analyzing a community and the steps needed to help design a sustainable community. In this important contribution to helping design sustainable communities, Mehrhoff, through his work with the Minnesota Design Team, seeks to "help communities take control of shaping a sustainable future of their own by means of information, insight, and civic dialogue." He urges readers to rethink the shape and shaping of their communities by looking at the idea of community in a more holistic and multidisciplinary manner. Mehrhoff tackles such topics as defining community, understanding the history of a community, understanding the issues and problems affecting a community, examining the visual aspects of a community, and obtaining citizen opinion throughout the process of becoming a sustainable community. Small communities everywhere can replicate the process discussed in the book. This well-written and thought-provoking book provides a nice blending of theory and practice and should be useful to all students, academics, local policymakers, and citizens who are interested in creating a common sustainable vision for our communities.

Roger Caves
San Diego State University

Robert J. Waste
California State University at Sacramento

Margaret Wilder
University of Delaware

PROLOGUE

I have witnessed, with both amazement and alarm, the rapidly accelerating changes occurring in and to American communities since the Second World War. As I've grown from wide-eyed child to bifocaled academic, the shape and shaping of American communities has been and remains the polestar for my life and work. In a very real sense for me, the personal has become the political.

American communities like those I have known have been flooded by a tidal wave of social, economic, and political forces. Their natural environments, social networks, economic structures, and even self-images have been engulfed and often destroyed in this process. Although these so-called megatrends have often produced important economic, social, and cultural benefits—such as new products, global communications, trade opportunities, and sometimes even an enriched sense of our common humanity—they have also contributed to many unhealthy patterns of local community development that seriously threaten our well-being now and in the future.

Trends, however, are not necessarily destiny. Our futures can be shaped by informed choices made today based on our vision of the communities we want tomorrow. Academic researchers and design practitioners have over time created many of the research and planning elements needed to fashion a holistic (systems) approach to community design. Such an approach to community design, however, is no mere academic exercise. A community design process that addresses and integrates both natural and human needs in a thoughtful, participatory manner, while focusing on long-term health instead of simply promoting "growth," is now not just possible. In fact, such a process has become essential to reversing the decline of our communities and to transmitting our natural and cultural heritage to future generations. This book represents one contribution to this process of designing sustainable communities.

The American Dream

During my lifetime, I have both witnessed and experienced profound changes to some of the most basic forms of community found on the American landscape. These changes can only be described with a deepening sense of loss for their passing. Nostalgia, however, is not a particularly productive form of social analysis. The key questions involve unraveling and refashioning the meanings of those changes.

My early years were spent in an old, inner-city neighborhood on the north side of Saint Louis, Missouri, where both my parents' families had dwelled for generations (Figure P. 1). A classic urban neighborhood in the Jane Jacobs idiom, the near North Side was filled with family, friends, church steeples and school houses, familiar parks, and intriguing corner stores. As a child, I would sit with my grandmother peering out the second-floor window of our tenement building, counting the passing cars to learn my numbers while unknowingly watching the future taking shape.

The juggernaut named the Interstate Highway System soon steamrolled some of those parks and old neighborhoods, paving the way for many of those cars I had counted while relentlessly siphoning off the people and vitality of city neighborhoods. The American Dream of the Fifties lured millions of families to greener pastures, often quite literally. For example, our church in north Saint Louis used to hold its annual picnic in a farmer's hall and picnic grounds miles away from the city. Within the span of a decade, the farmer's hall and grounds were surrounded by new homes and shopping areas. Families like mine were attracted to what had recently been farmland by the prospect of new, single-family homes with garages, located in pleasant natural settings. Exciting new shopping centers, filled with the things one now viewed on television, seemed to spring up everywhere like some new bumper crop amid vast expanses of free parking that made it easy to take the whole family shopping. After all, gasoline cost less than 20 cents per gallon; who needed a corner grocery store?

Cheap gasoline also made it easy for families like mine to take to the open road. Regular travel along the Interstate and state highways crisscrossing the Midwest now highlighted our summers. Getting out into the countryside, seeing the USA, seemed to be everyone's goal. However, it became increasingly apparent that what we were seeing from the

Figure P.1. Family Photograph Taken March 5, 1950, in St. Louis Park

windows of the old Plymouth was beginning to look remarkably different from what we had expected to see. The surrounding countryside started filling up with filling stations, fast-food restaurants, and, I gradually noticed, more and more empty farmhouses where I earlier had recalled families living. The dream landscape did not seem as promising anymore.

Rethinking the Dream

Perhaps because of my personal experiences with some of the dramatic transformations occurring to American communities, especially involving their physical landscapes, I became highly receptive to emerging new ideas about how to create more attractive, humane settings while preserving valuable aspects of our natural and cultural heritage. Like many college students of my age, I found the first Earth Day in 1970 to be an intellectual watershed. Earth Day tapped into a collective sense that there was something fundamentally wrong about the way the American Dream was taking shape.

Earth Day made it seem to this young scholar as though America had become one large university, with information and debates about ecology, pollution, population trends, and myriad related topics flowing rapidly across the country through books, journal and magazine articles, formal seminars, and informal discussions. Images of oil spills, traffic-clogged highways, and belching smokestacks filled the popular media. President Richard Nixon and the U.S. Congress seriously debated sweeping new environmental laws and regulations. The press and television tapped into the spirit of the times, devoting considerable coverage to environmental issues, some of it highly theatrical but some actually quite complex and insightful.

One of the more thoughtful programs during this period was a public television special about Ian McHarg, a planner and landscape architect who forcefully challenged the existing order of business. His refreshing new approach to development depended on a clear understanding and appreciation of natural processes. Perhaps because he had emigrated from his native Scotland to America, he seemed to bring a lively, critical detachment and fresh eyes to the development process that was engulfing us, but that we could no longer understand. In no uncertain terms, McHarg brought home the foolishness of the relentless overturning of the land and the communities built on it. His alternative to further destruction involved respecting and working with the natural processes underlying all life, including human communities, and bringing the full range of human knowledge, the natural sciences, social research, and the creative dimension of the arts and humanities to bear on how we shaped our world. An alternative future now seemed possible.

Taking It to the Streets

These epiphany experiences in the early days of the environmental movement inaugurated my own odyssey into community design. It is a career I have followed as both a practitioner and a professor for well over two decades. While pursuing a graduate degree in urban affairs, I developed a master's thesis considering the role of citizen participation in environmental impact analysis. The City of Saint Louis at that time was preparing to construct its last link in the Interstate Highway System, a link that would have obliterated several older neighborhoods, including the near North Side where my parents had grown up and my grandmother and I had counted cars from the tenement window. Building on McHarg's ideas about design, I tried to demonstrate that this project ignored a whole range of legitimate environmental and human concerns in the decision-making process. Environmental impact analysis was still in its infancy at that time, especially regarding the application of social research to planning practice, but the thesis furthered my thinking about community design for future reference.

I continued to work for years in the field of community design, first as a community organizer in an old south Saint Louis neighborhood, then as a city planner involved with a variety of projects including housing and neighborhoods, transportation, historic preservation and downtown development. It became increasingly apparent to me that the problems of central cities such as the one in Saint Louis were not isolated phenomena but directly related to those abandoned farmhouses I remembered from my childhood travels and linked by a view of land as a consumer commodity. A doctorate in American studies eventually followed, focusing on material culture studies, especially American cultural attitudes toward the natural and built environments, a stint as a museum educator at the Jefferson National Expansion Memorial in Saint Louis, then finally a move into university teaching.

Ironically, the physical, economic, and social decline of the old North Side neighborhood eventually made it attractive enough for reinvestment on the part of the city and redevelopment companies. Despite considerable abandonment and demolition, new and renovated housing has sprung up like new shoots after a ferocious forest fire.

Back to the Land

The study of American cultural attitudes toward the natural environment has proven to be an excellent avenue and guide to addressing current student concerns about community design. Environmental problems and issues now appear to be the major impetus for current student interest in urban studies. A new generation of students, raised in those suburbs I watched taking shape, for the most part now views cities as foreign and alien places. "Saving the cities" lacks the same dramatic appeal for them that it held for an earlier generation of urbanists like myself.

However, I have learned that these same students are extremely concerned about the general decline of a sense of community and especially about the rapid transformation of farmland and open space in and around their suburban homes into a hybrid or mutant form that is neither urban nor rural. Bridging the gap between their generational experience and mine has become a key part of my evolving odyssey into community design. To help myself better understand the community issues affecting my students, I became involved with the community design work of the Minnesota Design Team.

The Minnesota Design Team as Community Laboratory

The Minnesota Design Team represents a working laboratory for the study of sustainable community design. The Design Team is a pro bono organization established in the early 1980s to provide community design assistance to small Minnesota communities that would otherwise have not received such help. In addition to stimulating valuable design projects in over seventy communities, the Design Team has also generated a wealth of case studies about the issues facing communities.

Minnesota Design Team was founded by a group of design professionals who wanted to serve small towns and to encourage greater awareness of the positive role of community design. Previously known as the Governor's Design Team, it has, from its origins, been a volunteer organization comprising design professionals who donate at least one extended weekend each year to help communities envision alternative futures; team leaders contribute many more hours in preparation for the visit. Team members come from a wide variety of design professions such as architecture and architectural history, landscape architecture, planning, and interior and graphic design, with additional support provided by other disciplines or fields such as anthropology or marine biology. A voluntary steering committee provides ongoing policy and administrative support for the work of the organization.

A Design Team visit involves months of preliminary preparation by team leaders and members of the participating communities. The team selects for visits only those communities who demonstrate broad-based support for inviting the team. The town or neighborhood provides base maps and other valuable data about its physical, economic, social, and cultural characteristics. The three-day design charrette[1] begins on a Thursday evening when team members arrive in the town, often greeted by a welcome banner spanning Main Street. Local families host members of the team that weekend, feeding them extraordinarily well for the work ahead. An army travels on its stomach, and the Design Team has traveled far and well.

Community, the team has learned over time, is a complex phenomenon possessing multiple meanings. Consequently, the team spends Fridays listening to and observing its host community through a number of different lenses. Citizens analyze their community through a variety of background briefing sessions, SWOT (Strengths, Weaknesses, Opportunities, Threats) analysis, visits to local schools and senior centers, and bus and walking tours of the community. Building on Christopher Alexander's pattern language of communal eating, Friday's activities include a community-wide potluck dinner and town meeting open to anyone.

At the Friday night town meeting, the Design Team employs a nominal group process, which it more folksily calls "democratic brainstorming." Democratic brainstorming uses anonymous responses to questions about key local issues as well as small group discussions to overcome some of the typical barriers to communication in small communities; it also encourages an open discussion of the full range of community issues, which is recorded and publicly displayed.

On Saturday, the Design Team synthesizes the mountain of information it has acquired through Friday's activities. The team attempts to integrate the mass of background information into a distinctive design framework that reflects the unique characteristics of the host community. Team members further develop the key elements of this design framework, such as entryways, Main Streets, circulation patterns, or regional connections, into a series of graphic images to be presented to the community at a Saturday night town meeting.

The Saturday night presentation culminates the weekend design charrette. It attempts to generate widespread community interest and commitment to some of the new design principles and projects that have emerged during the past few days. When the Saturday evening presentation proves successful, discussions about the ideas and recommendations in the presentation often continue well into the next morning.

On Sunday morning, team members join their host families for Sunday brunch, additional brainstorming, and networking before bidding their adopted families and community farewell. A follow-up visit within the next year offers the community an opportunity to assess its progress toward its shared vision.

Although the procedures of a Minnesota Design Team visit have been studied, played with, and rationalized during the past fifteen years in order to make the process itself more effective, each team and its host community possess their own "messy order." No two communities or teams are exactly alike, nor do they all share the same visions for the future. Discovering and giving form to that uniquely messy order, I have concluded, is what community design is all about.

The University as Community Designer

Universities offer excellent venues for studying the messy order of communities. To be a professor by definition means to put something forward as being true, to advance a thesis and program about one's field of study. My personal odyssey into community design has led me to profess the firm belief that our future well-being as a civilization requires fundamentally rethinking the shape and shaping of our communities. Such a rethinking involves bringing the knowledge and resources of the academic world into a much closer, better organized, working relationship with citizens and practitioners attempting to fashion more livable, healthy communities. This book represents one attempt to help bring about this rethinking of community design.

Three basic assumptions underlie this effort at rethinking community design. First, I believe that the academic profession possesses a profound responsibility to help promote and create healthy communities. In a very real sense, we are supported by our constituents in these communities to serve them as guides through the uncharted seas of the new global village, provided with the time and resources to chart new courses for others to follow if we serve our missions faithfully. We can accomplish this mission through our teaching, academic research, and applications of our findings to community service. This book attempts to synthesize all three elements of the mission of the university.

The second assumption involves the design of the university in relation to contemporary communities. As the old saying goes, if your only tool is a hammer, every problem is a nail. The single-focus lenses of academic disciplines, although valuable as heuristic tools, distort the appearance and nature of our communities. Communities cannot be dissected or pinned beneath a microscope. They are complex social systems and need to be studied within this living framework. Academicians need to acknowledge the limits as well as the strengths of their academic disciplines in order to serve their communities more effectively.

The third and final assumption about rethinking community design underlying this work involves the need to operationalize the systems approach more effectively. Reinvigorating and building on the 1969 National Environmental Policy Act (NEPA), especially its requirement for environmental impact analysis of major federal projects, offers an excellent starting point for this crucial task. This landmark legislation not only embodied the ethos of the environmental movement of the 1960s, it also provided a meaningful model for a holistic approach to community design by its explicit recognition that communities involve complex networks of relationships. In particular, NEPA called for the integrated application of the natural sciences, social research, and the design arts to the practice of decision making on environmental issues; it also urged citizen participation in the design process.

Such a rethinking of NEPA does not represent nostalgia for the lost glory days of the 1960s but rather vital preparation for the needs of the twenty-first century. Although seldom fully used and often trivialized into pro forma bureaucratic procedures, NEPA still represents one of the best starting points for revitalizing not just federal projects but contemporary community design practice at the grassroots level. It can help in this important challenge by drawing the research and experience of the last quarter century into a true systems approach necessary for genuine community design.

Purpose Statement and Overview

Helping communities take control of shaping a sustainable future of their own by means of information, insight, and civic dialogue is the fundamental task of community design and the underlying purpose of this book. This work attempts to bridge the too frequent gap between theory and practice and to strengthen both elements in the process. Good theory should also be practical.

The book is divided into two main sections. The first investigates the problems facing local communities caused by the rise of a global economy; it also attempts to demonstrate the pressing need for a systems approach to community design. The second part examines the principles and practices underlying such a systems approach to community design. This section draws heavily on case studies culled from my work during the past eight years with the Design Team. It also considers issues involving the implementation of successful community design, such as a prototype identifying key criteria for success.

Like my own life, this book remains a work in progress, an outgrowth of seeds planted many years ago that are still growing and evolving. I hope that readers of this book, especially the next generation of students, will come to regard it as a valuable starting or reference point for their own odyssey into community design.

W. Arthur Mehrhoff
St. Cloud State University

NOTE

1. Design students at the Ecole des Beaux Arts would typically work up until the last minute on their final projects. The Ecole would send a *charrette* (cart) to collect their projects, and many students would climb into the charrettes to continue working on their projects. The design charrette is intentionally a short, two-day intensive collaboration,

with design work being conducted right up until the time of its public presentation. The compressed nature of the charrette generates a great deal of synergy and creativity among the participants.

PART I

The Need For Community Design

CHAPTER 1

THE THIRD WAVE

A Changing American Landscape

Despite my personal interest in understanding and influencing the direction of contemporary community design, those technological, economic, and demographic forces that have so dramatically altered the American landscape over the past half-century clearly transcend the life of any single individual. Futurist Alvin Toffler (1980) has designated post-industrialization as the Third Wave of epochal human social change, following the great transformations first from a hunting-gathering way of life to agriculture and then from agriculture to the Industrial Revolution.

Although still fluid and vague in form, certain key characteristics seem to define post-industrialization. One is the increasing importance of so-called service industries, such as health care, in comparison to the goods-producing industries that dominated the Industrial Revolution. Second, information and knowledge technologies have assumed a role of vital importance. Third, the related mobility, or footlooseness, of manufacturing industries, made possible by mechanization and automation as well as information technology, has permitted the physical separation of many of the processes involved in production; companies can now place many of their operations in different places, where they can find the most favorable operating conditions and potential for profits.[1]

Like its historic predecessors, this Third Wave of human evolution has thoroughly engulfed the physical, social, and economic fabrics of small towns, suburbs, and center city neighborhoods alike. Entire communities and regions are being transformed at a rapid rate, often far beyond their ability to comprehend, much less control. As one rural resident (quoted in Doyle, 1992) remarked, "It's almost produced a rural mentality that's fearful. . . . There's an undertone here. Are we losing control? What's going to become of us? It's all real scary" (p. 20A).

Small towns and rural regions have especially suffered from the dual impacts of declining manufacturing activities and the crisis in American agriculture. These communities typically have very little control of the corporate decision-making process (Blakely, 1979, p. 34). Larger farms depend far less on human labor as they rationalize operations into fewer, mechanized units. The loss of farms and subsidiary farming operations like implement dealerships and feed stores ripples throughout the entire community, as stores and related employment activities dry up along with the farms, and the young people head for greener pastures.

A MINNESOTA DESIGN TEAM PERSPECTIVE

Since 1983, the Minnesota Design Team has assisted dozens of such local communities as they attempt to come to terms with the highly disorienting effects of this massive transformation occurring to community life and identity (Figure 1.1). These powerful effects now reach into literally every aspect of community life. Although each community visited by the Minnesota Design Team remains unique in terms of its physical setting, history, and culture, the litany of community design issues and concerns they cite has become a very familiar refrain. The issues typically cited by communities in their applications for a Minnesota Design Team visit include:

- the loss of large areas of farmland and open space at the edge of the town to suburban-style residential and commercial development;
- the closing of many long-standing shops and businesses in the community, especially along their traditional Main Street;
- an outmigration of young people from their hometown to regional centers or metropolitan areas, as they search for improved employment opportunities;
- rapidly rising costs of infrastructure and services, often associated with the new development occurring on the periphery of the community;
- increasing pollution problems such as water quality due to leaching landfills, agricultural activities, or residential runoff into the community's water supply; and

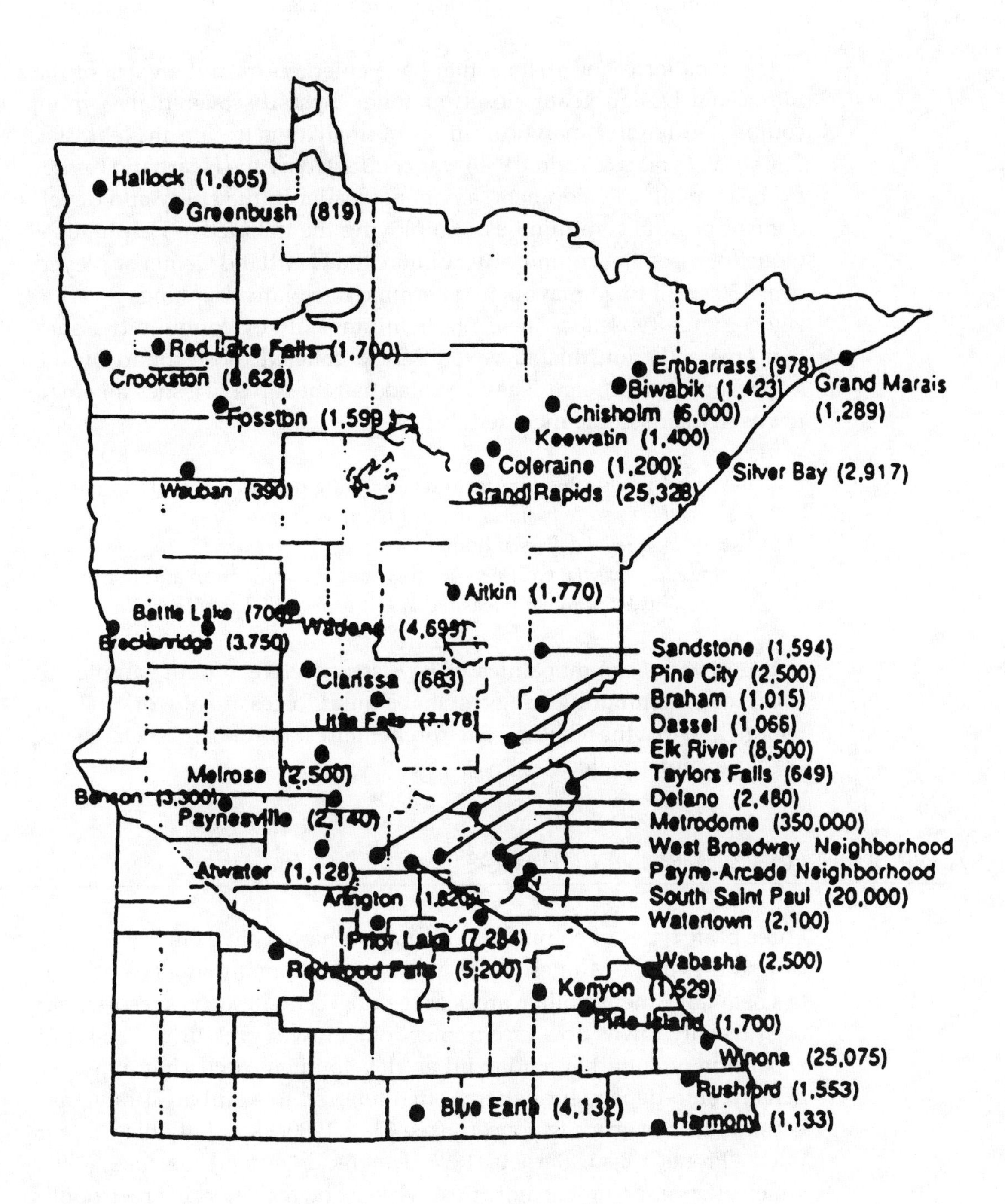

Figure 1.1. Communities That Have Received Visits from the Minnesota Design Team

- a general sense that the ties that historically bound community members to one another and to their physical site have deeply frayed and cannot support the new burdens increasingly being placed on the community.

These patterns and themes that have emerged from the work of the Minnesota Design Team closely parallel those discovered in a more comprehensive and analytical survey of small town leaders in Nebraska. The study (Kokes & Todd, 1990) was conducted by the Heartland Center for Leadership Development, a national leader in the study and development of rural communities.[2] Employing the interactive Delphi technique for assessing community opinion, the Heartland Center surveyed over 130 small town mayors to determine the relative rankings given to key community issues. The center's findings of key community issues that emerged from this survey closely paralleled those found by the Minnesota Design Team. They revealed that the top five issues for small towns in their sample included

1. new *employment opportunities* to retain young people in the community;
2. *expansion of existing business* and industry;
3. *keeping local retail dollars* at home;
4. *solving environmental problems* such as water contamination and pollution;
5. developing economically feasible *solutions to costly landfill problems.*

The obvious commonalities of key issues between both samples of small town communities suggest that similar forces are at work in both regions and that the resulting environmental, economic, and social issues are closely interrelated.[3]

A TYPOLOGY OF COMMUNITY CHANGE

Three basic types of community change scenarios have emerged in the case studies of the Minnesota Design Team. First, many small communities near large metropolitan areas or regional centers such as resort areas or university towns now face tremendous growth pressures. The rapid expansion of suburban and exurban development occurring in a low-density, auto-dependent pattern, often referred to as urban sprawl, has become their overriding concern (see Clay, 1980, especially his chapter titled "Fronts"; also, Barnett, 1996; Downs, 1994; and Garreau, 1991, which offers a somewhat more positive view on the emerging metropolitan form). Traditional farmlands and open spaces such as woodlands and wetlands surrounding the towns are rapidly being converted into expensive new subdivisions filled with enormous single-family residences, frequently built on an acre or more of land. This process also helps to fuel the development of related office complexes, fast-food franchises, and outlet malls on the metropolitan periphery. These new subdivisions are often built alongside traditional farming operations, with residents of these new subdivisions frequently complaining about the sights, sounds, and smells associated with agricultural activities. Communities caught up in this scenario don't seem to know what to make of this "progress." They have been told it's good for them, but the promise of new growth now seems greatly oversold as they frantically scramble to pay for new roads, public utilities, and schools while their citizens vocally demand lower taxes.

Declining small towns in outlying regions of the states represent a second community design scenario. A number of small towns, often based originally and almost exclusively on farming, mining, or the timber industry, have pleaded for assistance in stopping their downward free fall. These communities are particularly concerned about the loss of living-wage jobs, the rapid outmigration of their young people in search of more educational and employment opportunities, and the decline of their old Main Street commercial districts due to competition from shopping malls and megastores such as Wal-Mart.[4] This demographic pattern is by no means unique to Minnesota.

Several inner-city neighborhoods have also sought design assistance from the Minnesota Design Team to help rescue them from a tidal wave of change. In many key respects, their situations closely parallel those of the declining small towns in outlying regions. They, too, have lost jobs and residents due to the mass migration toward the metropolitan edge; their commercial districts resemble the forlorn Main Streets of many small towns in decline. In addition to these fundamental problems, however, urban neighborhoods like these must often cope with a complex overlay of racial issues not typically found in less ethnically diverse small rural towns. Furthermore, inner-city neighborhoods must frequently compete for scarce municipal dollars with powerful downtown financial centers and wealthier city neighborhoods seeking their shares of the city's tax revenues, making it even more difficult for these neighborhoods to cope with the cycle of abandonment and disinvestment.[5]

POST-INDUSTRIALIZATION: A SEA CHANGE IN COMMUNITY LIFE

All three types of communities perceive, either consciously or unconsciously, that a sea change has occurred in their everyday lives. Toffler's Third Wave is, to them, a very real phenomenon. According to economist Edward Blakely (1979), "We presently live at a time when the global society is undergoing total reformulation, creating an entirely new economic order linking communities to global economic conditions rather than to . . . self-regulating economic systems that characterized earlier [times]" (p. 308).

Post-industrialization, as the term clearly suggests, implies a quantum change from the industrial era and its corresponding physical and social arrangements (Castells, 1996; Drucker, 1986). During the 1980s, large numbers of American firms changed ownership, reduced operations, or altered their production processes. Whereas one out of three Americans were blue-collar workers in 1920, by 1980 only one in six made their living in such occupations. Drucker (1986, p. 776) estimates that the percentage of Americans engaged in manufacturing by 2010 will decrease to 10% of the workforce. The new global economy, whose leading producers are increasingly computer software developers, now depends much less on extracting large amounts of raw materials and far more on services and information technology.

Because of its shift away from raw materials extraction and processing, post-industrialization is generating an unprecedented restructuring of the global order of human settlements, including the structure and functions of local communities. Rural regions in virtually all developed nations have experienced a dramatic shift in their place within the larger system (Troughton, 1990, p. 23). In the famous words of novelist Thomas Wolfe, you can't go home again. Closed systems have been cracked open.

Certain key characteristics help identify post-industrialization as a qualitatively different phenomenon from the post-World War II industrial order led by the Big Three automakers, oil companies such as Standard Oil, and U.S. Steel. According to geographer Michael Troughton (1990), these key new characteristics include:

- rationalization into fewer large units;
- emphasis on high technology and education;
- declining requirements for unskilled and semiskilled labor;
- a shift from labor to capital; and
- continued migration to urban areas. (p. 24)

Not surprisingly, these strikingly new characteristics also place enormous stresses on the communities that had evolved to meet the needs of the previous industrial order. The question for community designers then becomes how to comprehend the relationships and patterns inherent in this new system and to apply them effectively to shaping future communities.[6] In Toffler's (1980) words, "So long as we think of them as isolated changes and miss [their] larger significance, we cannot design a coherent, effective response to them" (p. xx). To design such an effective response to these all-encompassing forces truly requires a systems approach to understanding human communities.

SYSTEMS THEORY: A FOUNDATION FOR COMMUNITY DESIGN

The systems approach to community design represents the confluence of several streams of thought that developed in different places in the World War II era. These intellectual streams include (a) the ecological model of the universe, such as the work of von Bertalanffy (1968), who first proposed the study of holistic systems as the "fulcrum" of modern scientific thought; (b) cybernetics (Bateson, 1972; Weiner, 1961), which added such concepts as input, output, and feedback; and (c) concepts of open systems and their interactions with their environments.

The systems approach was quickly found to be applicable to many diverse fields of study such as social systems (e.g., Buckley, 1967; Lewin, 1961) and to organizational management (Kast & Rosenzweig, 1970). Roland Warren (1963) devoted several chapters of his influential work, *The Community in America,* to considering the modern community as a complex social system. The systems approach has even been applied to mathematical models of the entire global environment (Forrester, 1970; Meadows, Meadows, Randers, & Behrens, 1972).

What exactly is a system? A system can be defined as any identifiable unit composed of at least two or more elements, held together by relationships that are integral and sufficient to the character and function of the system. One systems analyst defines a system as "an entity whose parts are seen to make up an orderly and complex totality, in accordance with some underlying set of rules" (Chetkow-Yanoov, 1992, p. 5). However, he further notes that this apparent orderliness of the system under consideration is really the product of the analyst's mind rather than an empirical fact. In other words, defining a system involves creating or defining the relationships. It is a heuristic and analytical tool.

The boundaries of any system can be defined at many levels to distinguish the system under consideration from its surrounding environment. In fact, systems analyst J. G. Miller (1955, 1978) claims that things or events ranging from the body to the family to communities to the universe, at all levels of complexity, can be viewed as types of systems. Following that same line of reasoning, Moe (1960) defined the local community as a "system of systems," ranging from families to international economies. The important point here for community design is that an observer needs to define the boundaries of the system under study without losing awareness of the particular system's relationships to larger systems. Around every circle, a larger circle can be drawn (Capra, 1975; Wolf, 1981).

The elements within any given system engage in recurring patterns of relationships through which they mutually influence each other.

Whereas *closed systems,* such as simple organisms, possess rigid boundaries that limit their interactions with their environments, *open systems,* such as human communities, are more fluid, exchanging energy, information, and resources with their environments (Olsen, 1968). Although they were always involved in larger systems of production and commerce, such as international commodity markets, many local communities often acted as though they were closed systems immune from larger forces in the world. Post-industrialization is rapidly breaking down those perceived walls.

Systems thinking further recognizes that open systems such as human societies involve multiple factors simultaneously operating on each other. It challenges linear thinking about cause and effect relationships, especially when dealing with complex systems such as human communities. "Not only do . . . many [factors] operate together . . . they also impact on each other as they interact. . . . Yesterday's outcome might well be a causal factor in the dynamics of what is happening today" (Chetkow-Yanoov, 1992, p. 130). The changing demographics of a community may result from changing economic relationships and processes of production, but the presence of newcomers to a community from a different ethnic group may also cause considerable change within that traditional community.

One criticism of the systems approach has been that it tends to emphasize equilibrium and "controls," often to the exclusion of conscious adaptations and the role of human values in decision making. People appear reduced to the level of algorithms. Several systems analysts have attempted to address this deterministic approach to understanding human communities. Boulding (1956) argued that human behavior ultimately depended on the image individuals and groups constructed of their worlds. However, he argued that these images of reality were not simply static but possessed a phenomenal capacity for growth and development, independent of external messages or feedback (p. 26). Lewin (1961) examined the concept of *steady state systems,* capable of altering their structures to adapt to change. Humans, according to Ackoff (1967) are *purposeful systems* who represent the interface between the organism and social systems. Buckley (1967) concludes that society is an *open system* responsive to its environment, complex in nature, and evolving new structures through a constant interplay with the environment. Amitai Etzioni (1968) defined an *active society* as one that not only planned for the future but also used control systems and feedback to guide societal development. Bogart (1980) also tried to show that feedback is only one form of strategic information exchange within a system. He also discussed the concept of "feedforward," strategic information that allows for images and memory, as well as the ability for communities to plan and anticipate future changes in the environment.

URBAN ECOLOGY AS A SYSTEMS APPROACH TO COMMUNITY

Communities struggling to come to terms with the implications of post-industrialization now need to engage in just such a thoughtful, purposive process. As anthropologist Gregory Bateson (1972) noted, "There is also latent in [systems theory] the means of achieving a new and perhaps more human outlook, a means of changing our philosophy of control and a means of seeing our own follies in wider perspective" (p. 477). At the turn of the twentieth century, an era as bewildering in many respects as today's rapid transformation, pioneering urbanists such as Robert Park and Patrick Geddes conceived what would today be called a systems approach of urban ecology to help urban scholars understand the dramatic effects of the Industrial Revolution on rural communities and cities. According to Park and his colleagues (1925, 1967), "The city is not . . . merely a physical mechanism and an artificial construction. It is involved in the vital processes of the people who compose it; it is a product of nature, and particularly of human nature" (p. 1). In effect, these early systems analysts argued that cities can be understood as complex, adaptive social systems.

Although formulated in response to the unprecedented conditions created by industrialization and the rise of large corporations, such as tenement slums and downtown skyscrapers, the urban ecology model can also help contemporary observers of communities better understand the post-industrial global economy as a dynamic, interrelated system involving five key, interrelated components:

- work and technology
- demographics
- environment
- organizations
- values

THE URBAN ECOLOGY OF POST-INDUSTRIALIZATION

The urban ecology model offers some valuable insights into the dynamics of this new post-industrial system. As with any complex ecological system, changes occurring in one part of the system (such as new telecommunications technologies and global corporations) alter the equilibrium between elements that evolved in response to other conditions. As previously noted, post-industrialization is proving especially destabilizing to traditional communities rooted in geographic locations based on raw materials extraction and production. Small towns tended to function as closed systems, somewhat self-contained in terms of their economic activities and often drastically limited in their social interactions. Today's small town, however, cannot escape the impact of world markets, demographic shifts, and environmental damage.

This traumatic shift occurs because the modern post-industrial corporation possesses a very utilitarian understanding of the local community. In the words of management expert Peter Drucker (1993b),

> The modern organization must be in a community but not of it. An organization's members live in a particular place, speak its language, send their children to its schools, vote, pay taxes, and need to feel at home there. Yet the organization cannot submerge itself in the community nor subordinate itself to the community's ends. (p. 7)

When the going gets tough, the organization simply goes elsewhere.

The primary reason for post-industrialization's ambivalence toward and destabilizing influence on local communities is because the new global economy depends much less than its industrial predecessor on the natural resource base of a particular physical environment and the demographic arrangements that evolved there, such as pools of unskilled labor, to work with those relatively scarce raw materials. Most of the costs associated with such prototypical post-industrial products as computer software or prescription drugs involve research knowledge and development rather than scarce materials. As Drucker (1986, p. 778) notes, the costs of prescription drugs are 50% knowledge whereas the manufacturing costs for semiconductor microchips are 70% knowledge. Movements of information and capital rather than raw materials have now become the main forces driving this new global economy; the size of the market for financial transactions such as currency exchange now far exceeds the market for industrial products. For example, world trade in goods and services amounts to about $3 trillion annually. By comparison, the London Eurodollar market handles twenty-five times that amount (Christenson & Robinson, 1989, p. 16). Information and capital are now the raw materials of the post-industrial economy.

The scale and organization of post-industrialization are also unprecedented and equally destabilizing to local communities. According to Warren Bennis (1993), "Global corporations have become the very models of postbureaucratic organizations, able to orchestrate a worldwide network of component units skilled at exploiting the specific realities of their local communities" (p. xiii). Many global corporations possess branches and subsidiaries throughout the world, giving them the ability needed to move resources quickly in response to rapid changes in market conditions (Flora, Spears, Swanson, with Lapping & Weinberg, 1992, pp. 144-145). Stockholders and board members of the new global corporations also reside all over the world. The loose networks of shareholders who make up these new global organizations lack allegiance to particular localities, making disinvestment in these local communities much easier to rationalize when economic returns weaken. Aided by new computer technologies and information networks in their global chess match, these international organizations continuously move their pieces to the best locations in terms of lower business costs, such as labor and tax situations, and ease of transportation to bring materials in and products to market. The post-industrial economy has become a network of information and capital flows instead of a structure of geographic places.[7]

The highly fluid nature of this new global system creates considerable stress on the equilibrium of local communities that had evolved in response to much different requirements during an earlier era of development. Essentially, each system, global capital and local community, possesses fundamentally different requirements and seeks fundamentally different goals. In the words of one observer (Gunn, 1991),

> Capital wants profits; communities want development. Communities want well-paying jobs for their residents; investors are driven to pay the lowest possible wages . . . at given levels of productivity. Capital seeks an environment free of costly regulation; communities require a life-sustaining ecology. Communities are defined by place and stability; capital is concerned with location primarily as a factor in transportation and transaction costs. (p. 2)

Not surprisingly, local communities generally feel overmatched in this life-or-death competition.

FEED FORWARD TO THE FUTURE

For some communities, the tensions have become unbearable. Even beyond immediate community concerns about the impact of physical and economic development, the Minnesota Design Team has discovered there now exists a more generalized anxiety about the loss of control and erosion of the community's character and quality of life. One observer (Luke, 1993) effectively captures the essence of this malaise:

> Community becomes . . . thin because workplace and residence, production and consumption, identity and interests, administration and allocation are so divided in an advanced industrial society predicated primarily on geographic and social mobility. This division of interests, loss of common historical consciousness, weakening of shared beliefs, and lessening

> of ecological responsibility is what necessitates alternative approaches to understanding community. (pp. 209-210)

Regardless of whether they are growing or declining, urban, suburban, or rural, these communities are now being forced to come to terms with the powerful forces of post-industrial change and to consider alternative approaches to thinking about what's going to become of them. They are being forced to design their own futures.

NOTES

1. The reasons for these shifts and their effects on communities are discussed in more depth in Parts I and II of Sternlieb and Hughes (1975).
2. "Gauging Community Opinion" (Chapter 6) contains a fuller description of the Delphi technique and a variety of other research methods for assessing community needs and opinions.
3. The closely interrelated character of these three elements is one of the central premises of the concept of sustainable development. This concept will be developed more fully in Chapters 3 and 7.
4. See especially Part Two, "Economy and Society," in *Rural Communities: Legacy and Change* (Flora et al., 1992, pp. 107-156) for a detailed examination of the effects of post-industrialization on rural communities. Flora is widely regarded as one of the leading scholars on the subject of rural America, and *Rural Communities* is one of the definitive works on the subject.
5. Suzanne Keller's *The Urban Neighborhood* (1968) is still a standard reference work on this topic. See also Hunter (1979), Wellman and Leighton (1979), and Melvin (1985). Majka and Donnelly (1988) offer a hopeful glimpse of several neighborhoods that have maintained a high level of cohesion despite their changing racial compositions.
6. The work of planning theorist John Friedmann created much of the foundation for the systematic analysis of the implications of post-industrialization for planning and community design. His *Retracking America: A Theory of Transactive Planning* (1973), especially Chapter 4, provided one of the intellectual foundations for this work.
7. The work of Manuel Castells deals in depth with the role of information technology in economic restructuring and the world order of cities. See *The Informational City* (1989) for one example of his analysis of this phenomenon.

CHAPTER 2

COMMUNITY

A Wave or a Particle?

And the people who lived in the towns were to each other like members of a great family . . . a kind of invisible roof, beneath which everyone lived, spread itself over each town. Beneath the roof boys and girls were born, grew up, quarreled, fought and formed friendships with their fellows, were introduced into the mysteries of love, married and became the fathers and mothers of children, grew old, sickened and died. Within the invisible circle and under the great roof everyone knew his neighbor and was known to him. Strangers did not come and go swiftly and mysteriously, and there was no constant and confusing fear of machinery and of new projects underfoot. For the moment mankind seemed about to take the time to understand itself.

Sherwood Anderson, *Winesburg, Ohio*

A WAVE OR A PARTICLE? LOOKING AT COMMUNITY IN NEW WAYS

Does community still exist in post-industrial America? Does the concept of community itself simply reflect a nostalgic longing for some bygone city neighborhoods and small towns, "good old days" that never were? As American Studies scholar James Robertson (1980) notes, "For Americans who are bewildered, bruised, or defeated by the . . . competition and loneliness of the modern world, the images of static rural community still offer refuge" (p. 240). Or is being part of a local community an essential component of human happiness? However nostalgic it may be, an archetypal concept such as community obviously speaks to deep American cultural needs and therefore should not be discarded lightly just for the sake of an undefined progress. The basic questions for community design to address then become: What do we mean by the term community? Can we afford to abandon this elusive yet enduring concept without causing terrible consequences for ourselves and future generations? If not, how can it be reconstructed to meet current needs?

Community is an important concept at several levels. "Thus we find two interrelated developments," writes community scholar Roland Warren (1972). "One is the actual change taking place in communities; the other is the change taking place in theoretical formulations among students of the community" (pp. 2-3). Considering the actual changes occurring in communities is vital to successful community design, but understanding changes in the concept of community may be equally important.

From the onset of the twentieth century, many scholars have fiercely debated the crucial theoretical concept of community from a variety of perspectives. The classic essay of Louis Wirth (1938) on "Urbanism as a Way of Life" did much to frame the terms of the debate about community in modern urban America. According to Wirth, urbanization caused qualitative changes in how people perceived and interacted with one another because of (a) increased population, (b) greater density, and (c) more diversity found in urban settings. From his perspective, urbanization (and now, by extension, post-industrialization) inevitably meant the disappearance of traditional communities.

Not everyone agreed with Wirth's sense of decline and loss of community. Other leading scholars on the subject of community, such as Herbert Gans (1962), argued that neighborhood and extended kinship systems have continued to exist despite growing urbanization, especially in urban ethnic neighborhoods, working-class suburbs, and small towns. Still others, such as Claude Fischer (1977; see also Cox, 1966), regard the concept of community as liberated by the forces of urbanization. Primary ties such as kinship are now spatially dispersed from the days when extended families lived in close proximity to one another; one's in-laws, thankfully or regrettably, no longer live down the block.

For better or worse, individuals are now increasingly free to choose the settings of their own communities based on similar tastes and lifestyles. One urban scholar (Fishman, 1990) calls this new type of community a "city à la carte" (p. 14). Much of the answer to the question of whether community still exists depends, then, on how one defines and operationalizes this crucial concept. The concept of community has historically been and continues to be the subject of considerable debate, because the debate about what makes up the good community is really a debate about what constitutes the good life (Filipovitch, 1989, p. 27).

LOST COMMUNITY

Community, however, is not just a theoretical construct. A sense of community also affects how well people live and work together in shared locations, as well as their future ability to cope with the effects of post-industrialization. The experience of the Minnesota Design Team proves instructive in this regard. Because it functions entirely as a volunteer organization, the Minnesota Design Team simply cannot visit every small community or city neighborhood that requests a visit. It therefore requires communities to formally apply for a Design Team visit and then selects two communities for visits each fall and spring.

Over time, the team has clearly recognized that communities demonstrating a strong sense of identity, meaning high levels of civic pride, involvement, and commitment by a variety of community organizations, typically benefit more from a Design Team visit than those where such broad-based support was initially lacking. Communities are now required to complete an application form that includes letters of support and financial sponsorship from a wide variety of community organizations, such as elected officials, civic organizations, business associations, and school leaders. Applications coming solely from City Councils or Chambers of Commerce are typically viewed with great skepticism and rejected as inappropriate to the task of community building.

Concern for the type of community described by Sherwood Anderson still resonates with many people. Randall Arendt and colleagues (1994) observed that "professional planners throughout the country are beginning to feel the effects of citizens' initiatives based on a growing public awareness that the special qualities of their small towns are being needlessly eroded by conventional sprawl development" (p. 25). Because community identity represents such a critical factor in the long-term success of a Design Team visit, applicants are now asked to respond to a series of questions (Figure 2.1) dealing with major community issues and concerns to assess their concept of community. One of the sample questions asks applicants to briefly describe the three most important problems facing their community. In response to this key question, many of the applicants have cited preservation of a sense of community as one of their major reasons for inviting the Design Team. For example, the City of Lake Elmo mentioned helping to promote new forms of development

Please respond to each question listed below. Attach a separate sheet of paper with your responses.

1. How did you hear about the Minnesota Design Team?

2. What do you think a Minnesota Design Team visit can do for your community at this time?

3. List and briefly describe the three most important problems for your community today.

4. List and briefly describe the three best opportunities for your community today.

5. List and briefly describe the three most important problems you believe your community will face ten years from now.

6. What do residents want the community and surrounding area to look like ten years from now?

7. Describe projects the Minnesota Design Team will be asked to address during a visit (buildings, landscapes, streetscapes, planning). How do these projects relate to the Design Team's purpose and your community's vision?

8. How do you plan to publicize and review the results of a Design Team visit?

9. What group will be responsible for coordinating the follow-through and implementation of ideas generated during the visit? What related experience do they have, and what is their understanding of your community's vision?

10. What are your first- and second-choice dates for a Minnesota Design Team visit? Remember that the visit begins on a Thursday evening and ends Sunday morning. It is important not to have any conflicts at that time, such as major community or school events. Why did you select these dates?

Figure 2.1. Minnesota Design Team Visit Application: Short Answer Questions

that would reflect its highly valued rural village atmosphere, whereas the City of Saint Joseph sought to prevent encroachment from the rapidly growing corridor along Interstate 94, so that its small-town atmosphere could be preserved. But what exactly does a sense of community such as a "small-town atmosphere" really mean in an urban, post-industrial American society?[1]

These typical responses by Design Team communities illustrate a powerful trend that is affecting many other communities in contemporary American society. A statewide focus group (Minnesota Planning Agency, 1993) revealed that major concerns among small towns include the feeling that (a) their fate is being determined by large organizations far away from the local community and (b) a pervasive fear of the loss of community control and identity. Bruce Hafen, a law professor at Brigham Young University, writes (in Leo, 1993) that "we are witnessing a gradual decline in the legal and social significance of community interests" (p. 31). Despite widespread resistance, a fundamental shift is occurring in American communities. In the telling phrase of Harvard political scientist Robert Putnam (1995), far too many Americans are now bowling alone.

TOWARD A WORKING DEFINITION OF COMMUNITY

By the term *community,* applicants to the Minnesota Design Team generally mean something much closer to Sherwood Anderson's traditional concept of a place-based community like Winesburg, Ohio, instead of a new network of computer users linked only in cyberspace by their common interest in *Star Trek* or some other phenomenon.[2] These traditional communities are physical places where basic human needs are met and primary relationships maintained. In most cases, they involve shared language and customs, as well as living closely to one another. Traditional small towns and urban ethnic neighborhoods represent prototypes of such communities.

However, "an adequate description [of community]," writes Warren (1963), "must somehow relate the community meaningfully to the rest of society" (p. 7). The dynamic forces of post-industrialization have exerted powerful effects on all the interrelated elements of the system of community ecology, so that traditional understandings of this key concept no longer adequately reflect current realities. Effective community design must begin by first considering and then coming to terms with the larger changes affecting traditional communities. We must take the time to once again understand ourselves.

COMMUNITY AS GROUPS OF PEOPLE

As the urban ecology model of Robert Park and others suggests, the traditional concept of community involves several key interrelated components that function collectively as a system. First, the concept of community always implies groups of people and their demographic characteristics. Historically, the groups of people living together in communities have shared many of the same demographic characteristics. These frequently included racial and ethnic traits as well as close kinship ties. Not only is it a place where everybody knows your name; many of them may indeed share it. Many of the small towns visited by the Minnesota Design Team have a very high percentage of their population who share the same ethnic background, such as being of Finnish, Norwegian or German descent. A person can move to such a traditional community, live there for decades, and still be perceived as an outsider by the native-born residents.[3]

However, the demographic characteristics of most communities are now in transition. Many traditional communities are now experiencing an influx of more diverse populations, perhaps affluent suburbanites drawn there by a desire to escape urban problems in the countryside or migrant workers simply seeking employment opportunities in a new factory recently built on the outskirts of the town. Geographer John Fraser Hart (1992) identified several types of rural communities that have been experiencing population increases. These include

1. towns that have four-year colleges;
2. towns on the suburban/exurban fringe; and
3. lake, resort, and retirement communities.

As the Minnesota Design Team has discovered over time, these demographic changes typically cause strong tensions within the communities about values and visions, tensions between growth and stability, tradition and change. Whose community is it, anyway?

COMMUNITY AS SOCIAL INTERACTION

Second, community has typically been characterized by a high level of social interaction. For many people, community life exists when one can go to a given location at any time of the day and see many people one knows by name. Those groups of people in communities are typically engaged in recurring networks or institutional relationships with one another, often based on kinship ties and a relatively simple division of labor. One shops at the local implement dealer or clothing merchant because that's how things have been done for generations.

Traditional communities know their members in different ways than others. There is a depth to their social relationships that can be either supportive or suffocating. A person buying food at the local store is not just a customer but someone's niece or the Lutheran girl or a member of the state tournament-winning basketball team. These social relationships are reinforced and deepened over time by annual community events such as fishing openers and festivals like Spud Daze, regular or chance meetings at the same local places like the Black and White Cafe, celebrations, and tragedies.

However, the social interactions that characterized traditional communities are now seriously challenged by post-industrialization. Warren (1963, p. 5) identified several key factors affecting the quality of community social interaction:

- increasing specialization of labor
- extension of the market economy (e.g., Wal-Mart)
- increased influence of government bureaucracy
- more differentiated interests (e.g., cable TV)
- growth of metropolitan regions

Although there are many economic and social advantages associated with these changes, such factors are often accompanied by the rise of "communities of limited liability." People with the means to do so increasingly choose where they live on the basis of lifestyle preferences, the race and ethnic characteristics of the neighbors, and an attractive physical setting. Such communities of limited liability are often characterized by the lack of close social relationships with immediate neighbors and the greater importance of broad social networks made possible by travel and telecommunications. The lack of "that hometown feeling" now seriously threatens the traditional community.

COMMUNITY AS SHARED VALUES

Third, demographics and social interactions contributed to a common outlook and values in traditional communities. Local religious institutions such as a church or parish frequently played a major role in shaping these values, encouraging cooperation and voluntarism or perhaps reinforcing a stern morality. Many a small town or urban neighborhood is identified by a church steeple or even an abundance of steeples. Ethnicity can also be an important factor in shaping community values. For example, the northeastern mining town of Embarrass, Minnesota, drew considerable strength from its traditional Finnish value of *sisu*, relentless determination, to help it rebuild itself from the impacts of a drastic downsizing of local mining operations. For its sense of community, the central Minnesota town of St. Joseph draws heavily for its identity on its Bavarian German background and the monastic orders that guide the two local universities.

However, as post-industrialization alters key elements of community ecology such as demographics and local employment patterns, social institutions and binding values are also affected. Many small communities no longer sense the solidarity, however imagined, of earlier times. Some scholars dating back to Max Weber at the turn of the century have argued that small towns and rural areas really have no separate existence anymore. Farm production, for example, is now organized in relation to commodity reports and government regulations; local political decisions are now oriented to state and federal programs regarding landfills or some other pressing local issue. To a great extent, the values and actions of local communities are increasingly determined by the agendas of distant organizations, whether they consist of television programs and movies influencing clothing styles of their young people or federal farm policy and international commodity exchanges determining what crops to plant and when.

COMMUNITY AS SHARED TERRITORY

Finally, community has traditionally implied a shared territory. For many traditional communities, such as native American tribes or the Amish, the people and the land are one and inseparable. Settlements such as small towns and urban neighborhoods often developed around prime farmlands, mineral deposits, timber, bodies of water, or even factories or docks. These physical features were integral elements of the cultures of these communities and gave them a sense of place, like the Mississippi River in the urban setting of South Saint Paul, Minnesota, or the falls in Little Falls, Minnesota. For example, Arendt et al. (1994, p. 4) identified several key geographic characteristics of the traditional small town:

- compactness of urban form
- medium density
- mixed-use town centers
- pedestrian-friendly
- rural open space at the edges

Now, the vast interstate highway network and a global economy have made many of these places functionally obsolete or simply one of many interchangable parts. Farms and open spaces are increasingly filled with residential and commercial developments. But the land is no longer rural in a true sense. Although it may still be used for agriculture purposes, it has become a part of the spreading metropolis. Small farms and Main Streets have now become endangered species. Franchises line the highway interchanges, making one town or neighborhood look much like any other.

Other shifts have likewise undermined a sense of shared community territory. Residents may live in the town or neighborhood but work elsewhere or have children in a geographically distant school district. According to many civic leaders, the fact that people don't work in the communities they live in any more makes time commitment a problem. It becomes harder to balance work, family, and a civic life. It's difficult to know one's place anymore.

REDISCOVERING COMMUNITY

In the words of architectural historian Spiro Kostof (1987), "Main Street . . . is much more than a place name to Americans. It is a state of mind, a set of values. It is what defined the heartland of the nation for generations" (p. 165). The place names of historic American towns such as Athens, Georgia, Syracuse, New York, and Cincinnati, Ohio, indicate that American culture, its origins grounded in the thought of classical Greece and Rome, has historically insisted that being part of the life of a community is essential to the good life itself. Although post-industrialization has seriously undermined the traditional definition of community, the need for community persists.

COMMUNITY AS KEY TO PERSONAL IDENTITY

Considerable support still exists for that historic American belief. The need for community is far more than a theoretical concern. In one classic study, psychologist Marc Fried (1963) found that grieving for a lost home and community was a serious and widespread phenomenon that occurred in the wake of urban dislocation. Although done in response to the planned dislocation associated with urban renewal programs, his study raises the issue of whether unplanned dislocation has a similar effect on residents of traditional communities. One scholar (Shore, 1993) argues,

> Without a sense of identity, a person cannot be whole; it is community that provides a man with his name. The social relationships, the responsibilities, the larger values, all help us know who we are. In an ideal community, a person's place would be so clearly defined as to make him indispensable. (p. 4)

Even the venerable *World Book Encyclopedia* (1995, p. 898) regards community as second only to family among the most basic human institutions.

COMMUNITY AND THE SOCIAL ORDER

Community provides important social benefits as well as psychological ones. Businessman Paul Hawken (quoted in van der Ryn & Calthorpe, 1986, p. 120) has suggested that when an economy slows down or even contracts, our lives come to depend much more heavily on the character of our communities. The loss of community, then, can have serious social and cultural consequences. The malaise of lost community also has implications for American politics. Daniel Kemmis (1993), the mayor of Missoula, Montana, maintains that

> people will respond to a politics that addresses their sense of what a good city or a good community might be, and how we would have to treat each other if we were going to go about the task of creating it. (p. 284)

COMMUNITY AS SOCIAL SYSTEM

Acquiring a human face in one's community has become exceedingly difficult in a post-industrial society. What sociologist Ray Oldenburg (1989) termed the "great good places," such as township meeting halls and neighborhood barbershops, have undergone tremendous changes and may even seem to have disappeared (see also Kemmis, 1993). However, the post-industrial revolution that has undermined so much of the ecology of traditional communities also offers some valuable theoretical insights that can help reconstruct the crucial concept of community.

Systems theory recognizes that all parts of a system are linked to and affect one another; however, it also suggests that systems can be subdivided into smaller, more manageable fields for study and operation. Although small towns and neighborhoods may no longer enjoy the relative isolation or independence they once had, they can still be perceived as meaningful fields of action if people so choose. Bennis (1993) suggests that organizations such as communities "are to be viewed as 'open systems' defined by their primary task or mission and encountering boundary conditions that are rapidly changing their characteristics" (p. 53). As always, the primary mission of communities is to endure and to support their inhabitants in meaningful ways, both internally and in relation to the external world.

Does community still exist in a post-industrial world with boundary conditions changing so rapidly? The concept of community ultimately depends on the level of analysis of the observer. According to Roland Warren (1963), "The community remains elusive, often encompassing one area and one group of people if looked at in a certain way, a different area and different people if looked at in another" (p. 8). People may look to their physical community to meet certain needs, but to broader networks for other important forms of social interaction. Community, then, is in the eye of the beholder.

What community designers need, therefore, is a model of community that recognizes and accounts for its dynamic, fluid qualities. Systems theory again proves highly useful in this regard. For example, modern quantum physics insists that energy in the form of light is simultaneously both a particle and a wave, not just one or the other (Wolf, 1981, p. 65). Just as light can be perceived as either wave or particle depending on the point of view of the observer, so we also need to stop thinking of community in either/or terms but rather in evolutionary terms as a complex and dynamic social system interacting with many other systems. Organizational analyst Sally Helgeson (1995) observes,

> The web provides a perfect metaphor for how science now perceives our universe in operation: not as a precisely calibrated great machine but rather as pulses of energy that continually evolve and assume shifting shapes as the various elements interact, and in which identity is inseparable from relationship. (p. 16)

COMMUNITY AS FIELD OF ACTION

Field theory is one application of systems theory that can be particularly valuable to the understanding and subsequent development of healthy, dynamic communities. It offers a much-needed holistic approach to understanding the systematic relationships between a place-based community and its larger social systems. By definition, a field is an unbounded whole with no clear boundaries, possessing only essential core properties or characteristics. Field theory helps a community designer understand a local community as a locally oriented social field operating within a variety of larger systems. Such systems include demographic shifts, information and resource flows, and the exchange of ideas. From the perspective of field theory, community can be viewed as simultaneously a place and a process, possessing both enduring core characteristics but also highly permeable boundaries where it relates to larger systems.

THE ECOLOGY OF HEALTHY COMMUNITIES

Building on systems theory, particularly the urban ecology model, the concept of feedback, and field theory, communities can now also be regarded as adaptive organisms and social systems that continually respond to changes in their external and internal environments (Luther, 1990, p. 44). Like all living organisms, they will attempt to regulate themselves to achieve a satisfactory balance with their external environments, finding their proper niche within the larger ecology of relationships. This balance will always be a dynamic one, requiring constant monitoring and adaptations to new stimuli.

A healthy community system, concludes Dykeman (1993), is "continually creating and improving those physical and social environments and expanding those community resources which enable people to support each other in performing all the functions of life and in developing themselves to their maximum potential" (p. 5). Although community in its traditional form as a closed system may no longer exist, it is still needed and therefore must be re-created in a new, more appropriate form. After all, everyone must be someplace.

NOTES

1. The work of Arthur Vidich and Joseph Bensman (1968) on small towns in mass society is especially valuable in understanding the impacts of urbanization and post-industrialization on rural regions.
2. See the March-April 1995 issue of *Utne Reader* for an extended debate on "Cyberhood vs. Neighborhood." The issue features numerous articles dealing with both the positive and negative impacts on communities associated with the new telecommunications technologies such as e-mail and Internet, as well as considering the characteristics of place-based communities.
3. The stories of Minnesota writer Garrison Keillor effectively communicate the humor, pathos, and ambivalence of traditional communities in a way no scholarly work can hope to. See *Lake Woebegon Days* (Keillor, 1990) for an introduction to this imaginary but very real world.

CHAPTER 3

IS THERE A PLACE FOR PLACES IN COMMUNITY?

For several decades now, landscape architect and journalist Grady Clay (1973, 1987, 1994) has closely observed and interpreted the changes occurring to American cities. His thoughtful insights about the meanings of the built environment offer valuable clues to the emerging post-industrial order. Like Robert Park, another urban studies pioneer, Clay employs a systems approach to the study of urban environments. He regards the city as an organism that is constantly in motion, always evolving in response to new forces such as technology or demographic shifts.

For Clay, the city is a place-process. "Cities," Clay (1980) writes, "are forever rewriting their repertoires" (p. 11). Although cities may appear solid on their surfaces, underlying forces, like the geological processes of plate tectonics, are causing massive shifts and major disruptions to existing urban patterns. Clay looks carefully at the built environment to uncover visual evidence of these tectonic shifts in how human settlements are organized. He uncovers visual clues to urban patterns, such as changes in the types of raw materials stacked along transportation routes or zoning hearing notices at the edge of farmlands, to help decipher what political and economic forces are acting on the communities. One particularly telling example involves the closing and abandonment of grain elevators along many small town Main Streets, a phenomenon the Minnesota Design Team has often observed.

Figure 3.1. Farmland Destined for Future Development
Photograph by James R. Dean. Used with permission.

Visual clues such as these can offer community designers important insights into the main characteristics of the evolving post-industrial city. One of its main characteristics is its relentless transformation of farms and open spaces at the city's boundaries into new urban developments.

The rapid replacement of farmsteads with franchises along our highways indicates that the process of post-industrialization radically transforms the relationships between cities and nature, between economics and places. The transformation is not altogether a happy one. Clay (1980) concludes that "the story of the city is an account of how mankind has used new wealth and energy to exploit the natural world; the end of the story might describe the end of cities as we have known them" (p. 14).

With its relentless emphasis on instantaneous communications and assumptions about the interchangeability of people and goods, post-industrialization has severely undermined many traditional concepts and aspects of community. This process appears especially true regarding the importance of strong personal and group connections to particular geographic locations. Footloose global industries and corporations also seem to require easily uprooted physical facilities. Even professional sports franchises like the legendary Cleveland Browns, whose very identities used to embody the spirit of gritty urban places, increasingly appear adrift and disconnected from their communities. As many Minnesota Design Team communities have discovered to their shock and dismay, the new global economy places little value on the surrounding farms, woodlands, wetlands, or the historic landmarks of small towns and urban neighborhoods, except as inputs into the process of production and consumption.

COMMON COMMUNITY DESIGN ISSUES

Applicants to the Minnesota Design Team typically express deep-seated anxieties regarding what they perceive as a greatly weakened sense of community caused by rapid socioeconomic changes. However, they often cite very specific concerns about the physical characteristics and impacts of recent development as their primary reasons for undertaking a community design process. Like the canaries once used by coal miners to detect early hints of gas leaks in the mines, these growing citizen concerns indicate the presence of undetected but dangerous structural flaws in the post-industrial system.

These concerns, however, extend well beyond Minnesota Design Team communities. Empirical evidence on growth management efforts across the country indicates that such efforts are generally brought about by severe financial and environmental concerns rather than by elitist efforts to protect their privileged positions.[1] In *Our Common Future* (1987), the World Commission on Environment and Development states the matter even more simply. "The public," it concludes, "puts a value on nature that is beyond the normal economic imperatives" (p. 165).

Applications to the Minnesota Design Team do, however, provide an excellent sampling of these widespread environmental concerns. Key community issues involving the physical environment generally fall into two basic categories. First, many of the concerns they expressed deal with overall community appearance. Sometimes, this issue takes the form of abandoned storefronts on the town's or neighborhood's Main Street, as in Little Falls, Minnesota. Other communities, like Lake Elmo, watched the development of "prairie palaces," massive homes on several-acre lots, carve up the precious rural landscape around the town into small kingdoms. Still others, like Clearwater, Minnesota, puzzled at the maze of drive-in establishments lining the highway interchange to Interstate 94, creating a gaudy gateway to what was once a riverfront town. Grand Rapids regarded the loss of its remaining white pine trees near the city limits as an issue affecting community appearance, the natural environment, and the fundamental identity of the community itself.

Sometimes, however, these concerns extend much deeper, reaching into the underlying natural systems of the community. Paynesville residents expressed serious concerns about the impacts of lakeside development on overall water quality in the region. The porous karst soils in Lewiston meant that any runoff from farming operations in that southeastern Minnesota town was easily absorbed into the underlying aquifer. Large-scale pivot irrigation systems now employed in some farming communities visited by the Minnesota Design Team required cutting down shelter belts of trees and windbreaks planted during Dust Bowl days to allow the irrigation systems to turn more easily. Although removal of the shelter belts may allow greater crop yields, it also poses serious threats to the soil and the wildlife habitat that has evolved in them. In the complex system of urban ecology, "Doing your own thing" can often undo someone else's.

UNRAVELING THE SYSTEM OF PLACES

As the previous examples indicate, external forces are now placing enormous stresses on the geography of local communities. The American land use system in particular, predicated as it is on the autonomy of individual municipalities and representing the interactions of millions of individual decisions about the use of land on a daily basis, depends on an awareness of and respect for the ecology of specific localities for its very survival. Current evidence suggests that such awareness and respect may be sorely lacking.

Although the global orientation of post-industrialization has created some valuable new socioeconomic arrangements and outlooks, it has also contributed to a fundamental lack of attention toward natural systems in particular places. This "can't see the trees for the forest" attitude (sometimes even a "don't need the trees" philosophy) is demonstrated in a variety of ways in relation to shaping human settlements.

Buildings

For millenia, humans designed their buildings in accordance with the limitations of the particular site, available technologies, and resources (Rapaport, 1982). For example, desert peoples of the Islamic world used natural cooling systems and water to create an effective and distinctive architectural style. Plains Indians created the highly mobile and elegant tipi from buffalo skins and pine poles.

However, one of the main legacies of modern architecture is its construction of buildings that separate people on the inside from the natural world. These "machines for living" can be located anywhere in the world, making human settlements look increasingly alike. One of the main complaints typically put forward by communities applying for a Minnesota Design Team visit concerns the appearance of standardized franchises along the main roads leading to and from the communities. Meanwhile, building these structures often requires enormous amounts of nonrenewable natural resources as well as energy to internally heat and cool.[2]

At the same time, many historic or just plain old structures that still possess useful lives if properly maintained are being relentlessly demolished. These older structures, frequently built by local workers of materials found in the surrounding region, typically gave a community its unique physical and cultural character. Their steady disappearance represents not just the loss of the energy and resources needed to build them, but also many of the qualities that linked members of that community to that specific place.

The Costs of Sprawl

Urban sprawl, the pattern of low-density development reaching into farms and open spaces at the edges of metropolitan areas and totally dependent upon the automobile, also displays a fundamental disrespect for the limits of natural systems.[3] For example, urban sprawl greatly accelerates the amount and flow of storm water runoff due to its concentration of impervious surfaces such as roads, driveways, and parking lots. The total amount of storm water runoff from a one-acre parking lot represents sixteen times the amount of runoff from an undeveloped meadow. In addition, this storm water will carry large amounts of unfiltered pollutants such as motor oil directly to local storm sewers.

This development pattern disrupts more than just the hydrology of a region. It can dramatically affect systems of wildlife as well. Environmental psychologist Ralph Taylor (in Gallagher, 1993) reminds us that "even if only ten or fifteen acres out of a hundred are perturbed, that can mean a radical disruption of a habitat, because breaking into a core area of some species can make a whole territory dysfunctional" (p. 200).

Human systems themselves are not immune from the costs of sprawl. A recent Urban Land Institute literature review (Frank, 1989) found these costs to be between 40% and 400% higher in low-density regions, compared to those with more centrally located services and facilities. According to one comprehensive report (Chen, 1995, p. 3), municipal services and, indirectly, taxpayers, are especially strained to fund sprawling development patterns. In the same report, Jim MacKenzie and Roger Dower of the World Resources Institute estimated that the indirect costs of sprawl not borne by users could reach as high as $355.7 billion annually (cited in Camph, 1995).

As farmlands near metropolitan areas are developed into new subdivisions and commercial centers, agriculture becomes increasingly concentrated into larger units of production operated according to industrial models. Success in commercial agriculture is increasingly measured in terms of cost per unit of production. Agricultural efficiency is evaluated in terms of scale of production, revenues, and greater reliance on secondary inputs such as chemicals rather than human labor (Troughton, 1990, p. 25). Many communities in Minnesota and in other states now find themselves wrestling with the serious environmental problems associated with large feedlots and hog farm operations, as well as with the economic and social impacts of declining farm populations. Meanwhile, vast amounts of valuable agricultural land disappear forever.

Unfortunately, this industrial model of commercial agriculture yields some thoroughly unintended and unpleasant results. At a recent symposium on the well-being of the Mississippi River held in Louisiana, hydrologists and other scientists warned that agricultural chemicals used in the Midwest were creating an expanding zone in the Gulf where aquatic life no longer existed ("Dead Zone," 1996, p. 2A). Findings from the U.S. Geological Survey indicated that most of the Mississippi River's nitrogen comes from agriculture operations. Large amounts of nitrogen fertilizer compounds from farms near the Mississippi River flow into the Gulf and stimulate algae growth faster than the ocean system can accommodate.

Global Trade Networks

Lack of attention to particular localities also reveals itself in terms of international trade within the new global system. The new system of international trade seriously undervalues natural resources by treating these limited materials as what economists might call "free goods." "Money has no ecological consciousness," concludes the *Gaia Atlas* (Myers, 1991). "If timber from a virgin tropical forest costs less than a sustainable source in Scandinavia [or Minnesota], then it will be hard to stop its extraction. Environmental damage is still not included in the price of goods and food" (p. 34).

Furthermore, depletion of nonrenewable natural resources in one region of the country or world can now be replaced by simply moving one's operations to another area. This constant movement of resources and products over long distances also makes recycling and remanufacturing of these materials more difficult. Many communities now find themselves dealing with abandoned quarries and overflowing landfills as evidence of this attitude toward particular locations. In these and many other key respects, the post-industrial system does not consider the environmental impacts of its economic actions.

How the Cake Crumbles

Despite its global reach and enormous achievements, however, post-industrial society rests on some very fragile foundations. Post-industrial attitudes toward nature and place are seriously flawed and highly destructive of the very sources of their existence and well-being. The mental separation of the built environment from the natural world that characterizes the post-industrial system underlies many of our current environmental problems.

The economic models that guide the creation of the global economy appear especially blind to the role of the natural environment in this new order, often regarding it as subordinate to economic requirements rather than a priori. In the words of landscape architect Ian McHarg (1969), "The components which the [economic] model excludes are the most important human ambitions and accomplishments and the requirements for survival" (p. 25). Entrepreneur Paul Hawken seconds McHarg's view. Hawken (1993) maintains that "in many ways business economics makes itself up as it progresses, and essentially lacks any guiding principles to relate it to such fundamental and critical concepts as evolution, biological diversity, carrying capacity, and the health of the commons" (p. 5).

Because of its large blind spot regarding the role of the natural environment in the total system of production, post-industrialization has become extremely destructive of its own source of wealth. The new market economy, observes Murray Bookchin, (1974, pp. 230-231) is replacing a complex organic environment with a simplified and inorganic one, literally disassembling a biosphere that has supported humanity for countless millenia in favor of a synthetic one.

One of the most fateful errors in the new global economic system is its inability to recognize that it consumes the very source of its existence. According to British economist E. F. Schumacher (cited in Henderson, 1988, p.176), it treats limited fossil fuels, natural systems and humans themselves as interest rather than as capital. Community activist Hazel Henderson (1988) compares the system of post-industrial production to a layer cake with icing. According to Henderson, the official market economy represents the icing on the cake (p. xviii). Its very survival depends, however, on the underlying layers consisting of the public sector, community support networks such as civic groups and volunteers, and, ultimately, the natural environment itself to supply all these resources and to absorb the waste products of the other layers. When those bottom systems of community and environment collapse from neglect and abuse, the entire structure crumbles. As the saying goes, you can't have your cake and eat it, too.

THE BREAKDOWN OF SHARED TERRITORY

Concern about the breakdown of shared territory as an integral component of community identity extends far beyond Minnesota Design Team communities. This issue of harmful urban development patterns has assumed not just local and regional but national and even global significance. The environmental impacts associated with the widespread breakdown of a sense of shared territory have affected community systems at all levels.

In addition to the local environmental concerns typified by Minnesota Design Team communities, current development patterns have emerged as serious problems at the state level in the United States. The Minnesota Environmental Quality Board (1993) issued a report examining the environmental impacts of population and development pressures throughout the state. The report identified several growing problems such as loss of forests and wetlands, increased urban water runoff, and water contamination from farming and on-site sewage systems. This report was followed by another (Sustainable Economic Development and Environmental Protection Task Force, 1995) encouraging more environmentally sound approaches to development.

At the state level, many other states besides Minnesota have identified similar concerns about community development patterns. Growth management authority John DeGrove (1996) observes that "experience in [many] states suggests that a state land-use framework is a critical ingredient in the state and regional management of growth" (p.1). The State of California, for decades the prototype of the sprawling American landscape, issued a report (cited in Chen, 1995) dealing with the state's future development. The report, jointly developed by a highly diverse coalition consisting of the State of California Resources Agency, Bank of America, the Low Income Housing Fund, and the Greenbelt Alliance, concluded that California's future could not be shaped successfully unless it moves beyond sprawl (Chen, 1995, p. 6).

There is also a consensus emerging at the national level in America that current development patterns are seriously flawed and highly destructive of community environments as well. One of the first efforts at assessing the environmental impacts of low density development was Real Estate Research Corporation's (1974) notable *The Costs of Sprawl.* That pioneering study concluded that sprawling development was generally the most expensive form of residential development in terms of economic costs, environmental costs, natural resource consumption, and many personal costs such as increased commuting time.

As previously noted, social researchers David Godschalk and David Brower (1989) examined a wide range of growth management efforts throughout the United States. They concluded,

> A powerful impetus for growth management was concern for the environmental impacts of growth. . . . We have become aware that this issue is much more complex and intractable than we originally believed. . . . The environmental effects of growth are not simply local or regional; they are global. (p. 173)

The President's Council on Sustainable Development (1996), a broad group consisting of business, citizen, and environmental interest groups, reached conclusions similar to those of Godschalk and Brower. The Council cited a number of environmental concerns about current development patterns. These included

- global deforestation;
- loss of biodiversity;
- declining fish harvests; and
- increasing global pollution.

Although the concept of sustainable development (as opposed to growth of consumption and output) is still ill-defined and heavily contested in policy circles, the Council concluded that "Americans' hopes for the future are linked to the rest of the world" (p. 5).

At the same time, important alarms are now being sounded at the international level about the environmental impacts of the emerging post-industrial global order. The first such alarm was the pioneering work entitled *The Limits to Growth* (Meadows et al., 1972). This study stressed the existence of environmental limits to unrestricted population and development pressures. Despite some serious criticisms of its research methodology and mathematical modeling techniques, its central thesis about the limits of natural systems has now become commonly accepted. For example, the book predicted that we would find pollution "sinks" on the Earth that would fill up as a result of unrestrained population and economic growth. Growing concerns about acid rain, expanding ozone holes, deforestation, and global warming all seem to support its solemn premise.

World leaders are now seriously considering the implications of limits to growth for future development patterns. The United Nations World Commission on Environment and Development (1987), widely known as The Brundtland Commission, helped popularize the concept of sustainable development. In its report entitled *Our Common Future,* the Commission argued that sustainable development must meet current social and economic needs without preventing future generations from meeting theirs. The Commission particularly focused on the huge imbalance of wealth and consumption between the industrialized nations and those of the developing world. Because developing nations avidly absorb popular media and copy the development patterns of industrialized nations, which have helped spawn the environmental crisis, the Commission maintained that these development patterns must be fundamentally altered. The Commission warned that thresholds exist that cannot be crossed without endangering the basic integrity of global systems and that today we are close to reaching many of those thresholds and endangering the survival of life on Earth.[4]

In Agenda 21, adopted at the United Nations Conference on Environment and Development in Rio de Janeiro in 1992, developed and devoloping nations agreed that environmental threats are becoming increasingly serious and that more sustainable development is needed (Yanshen, 1995, pp. 5-7). However, industrialized nations have taken a stronger stance on the need for environmental protection to preserve resources for future generations, whereas many of the developing nations regard grinding poverty as a form of pollution and emphasize more equity within the existing generation.

COMMUNITY DESIGN AS PLACE-ORIENTED PROCESS

These critiques of harmful development patterns all clearly suggest that the process of post-industrialization and its utilitarian approach to local communities exacts a heavy toll on the physical environment in terms of the health of individual communities and larger ecosystems as well. The stakes are enormous. In the words of Sir Patrick Geddes (1968), "From this [utilitarian] standpoint, the case for the conservation of Nature . . . must be stated more seriously and strongly than is customary. . . . On what grounds? In terms of the maintenance and development of life" (p. 94).

Geddes's statement further suggests that conservation of Nature, in the form of shared territory such as natural systems and public spaces, needs to remain a fundamental element of any new working definition of community. This new understanding of the relationship between human settlements and nature applies directly to community design.

Unfortunately, our current system of metropolitan governance clearly works against such an ecological approach. Pioneering urban scholar Norton Long (1962) conceived of urban policy making in American metropolitan regions as "an ecology of games" between competing interests such as businesses and labor or ethnic groups. This highly evocative image suggested that public policy making was not conducted on the basis of disinterested rationality but in reality constituted a dynamic system of often unplanned interactions involving a large number of such games or special interests.[5]

However, in this ecology of games, no players clearly represent a long-term concern for the public interest in the health of the region's natural environment. Few boundaries of political jurisdictions correspond to those of ecological systems such as watersheds, and many communities feel a strong need to compete with their neighbors for the increased land values caused by development. This lack of clear, long-term public interest and political representation in decision making especially affects the game of the natural environment as it is played out in the global economy and experienced in local communities. If the natural environment does represent the foundation of all the other games within the community and is indeed being seriously threatened by present development patterns, then what Long (1962) termed "a responsible ordering of the community" (p. 255) now requires treating the natural environment as the logical starting point for healthy community design.

Although much current environmental policy at all levels talks about balancing the needs of business with citizen demands for environmental protection, the environmental balance sheet is the ultimate measure of successful community design. "City and country are one thing, not two things," wrote Lewis Mumford (1968), "and if one is more fundamental than the other, it is the natural environment, not the man-made overlayer" (p. 169). Community prosperity in the long term must rest on healthy natural systems. In fact, a community's natural environment can become the logical starting point for undertaking a fundamental reordering of a community's identity and future vision.

An ecological systems understanding of community design is needed to effectively reorder our present development patterns. Ecology involves the study of the reciprocal relationships between organisms such as communities and their environments. This ecological systems approach also forces us to consider ourselves as organic members of the environments we inhabit rather than as detached from nature in our decision making.

As Geddes (1968) observed, real human wealth and health rest on the vitality of the supporting natural and community environments. Thomas Michael Power, chair of the Economics Department at the University of Montana, echoes Henderson's (1988) ecological understanding of economics. According to Power (1996),

> The social and natural environments should figure prominently in our economic view because they are the only sources of economic raw material. Natural resources flow from the natural environment and labor productivity flows from the social environment. . . . Moreover, the social

> and natural environments make life more meaningful, satisfying, and diverse. (p. 19)

Political scientist Meredith Ramsay (1996) reached a similar conclusion. According to Ramsay,

> Possibilities for the reclamation of [communities] lie not in economic development efforts primarily . . . but in the emergence of . . . new institutions and new civic cultures that foster a sense of responsibility for and commitment to the local community. (p. 114)

Communities that lack clear, detailed knowledge of their supporting natural and social environments now run the very likely risk of not adapting to important changes in these environments.

In particular, a strong sense of connectedness to geographic localities constitutes a logical and vital element of such an ecological understanding of community design. Planner John Friedmann (1993, p. 483) argues that localities and regions are the spaces of people's everyday lives, and those spaces remain very important to them, an assertion strongly borne out by interviews with people in Minnesota Design Team communities. Environmental psychologist Winifred Gallagher (1993) concludes that "we need places that support rather than fragment our lives, places that balance the hard, standardized, and cost-efficient with the natural, personal, and healthful" (p. 19). Ecologist Aldo Leopold (quoted in Bradley, 1996) argued that there has to be

> some force behind conservation more than universal profit, less awkward than government, less ephemeral than sport, something that reaches into all time and places where [people] live on land, something that brackets everything from rivers to raindrops, from whales to humming birds, from land estates to window boxes. I can only see one such force: a respect for land as an organism; a voluntary decency in land use exercised by every citizen and every landowner out of a sense of love for and obligation to that great biota we call America. (pp. C19-C20)

Community design is the force that Leopold so eloquently describes; localities and regions represent the playing fields where the daily drama unfolds. In Gallagher's (1993) words, "we must put the principles emerging from the multidisciplinary science of places into practice on local and global levels" (p. 19). Despite the tidal waves of global restructuring that are occurring, these remain the spaces of most people's everyday lives. As the Minnesota Design Team has discovered, the qualities of those spaces, particularly the common landscape of water, parks, open spaces, and civic places, remain practically and psychologically significant to their lives as well as increasingly important economically. Although people may not be able to roll back the tides of the new global order, they can clean up its beaches in their local communities through the voluntary decency in land use that is created through the process of community design.

NOTES

1. See a major study by David Brower, David R. Godschalk, and Douglas R. Porter (1989) on growth management strategies and issues from across the United States.

2. The Audubon Society developed a remarkable prototype of an environmentally sustainable building to demonstrate the concept of sustainability. The Society renovated a hundred-year-old department store in downtown Manhattan as its world headquarters, emphasizing recycled and environmentally benign materials as well as energy conservation, in order to demonstrate the feasibility and benefits of "building green."

3. See Brewster (1996). The Urban Land Institute, a professional organization representing the real estate development community, has taken a strong stand on the need for environmentally responsible development patterns. Brewster's article clearly articulates the environmental and economic benefits such development creates.

4. The literature on sustainable development is vast and rapidly expanding. Some valuable works include the important theoretical work of economist Herman Daly and John Cobb (1989), Lester Brown and Linda Starke (1994); and "Entrepreneurs and Ecosystems: Building Sustainable Economies," in the January 1996 issue of *Northwest Report*, pp. 1-7. In addition to the previous sources, the Center of Excellence for Sustainable Development was created by the U.S. Department of Energy's Office of Energy Efficiency and Renewable Energy. The Center's stated mission is to provide communities with the best information on sustainable development and to link people to relevant public and private programs promoting sustainable development. Through its website (*http://www.sustainable.doe.gov*), the Center provides access to articles on the topic of sustainable development as well as case studies of successful applications.

5. See Norton Long (1962) for an extended discussion of metropolitan governance. One of the most important games affecting community design is the so-called Growth Machine, or growth politics. A great deal has been written about this concept of related interests such as banking, real estate, sometimes labor, and elected officials who continually promote development at the expense of other community needs. See Logan and Molotch (1987) for a fuller discussion of this crucial concept.

PART II

Methods of Community Design

CHAPTER 4

ACTION RESEARCH

The Foundation of Community Design

As Part I of this book has indicated, traditional understandings of community can no longer be taken for granted in contemporary American society. To a considerable extent, then, the concept of community must not only be reevaluated but re-created for the new conditions of modern American life. Although to some people it may seem like reinventing the wheel, creating new visions of community needs to be done in order to move into a healthy future. Some existing models already point the way.

WITHOUT VISION THE PEOPLE PERISH

Portland, Oregon, immediately appeals to the first-time observer as a model for regional development. Flying over the region reveals definite boundaries to the spread of urban development and vast amounts of open space. Public transportation is technologically sophisticated, readily available, and user-friendly. Portland's downtown remains a vital and attractive destination, filled with lively plazas and pedestrian amenities such as a Lawrence Halprin fountain, which echoes the nearby Cascade Mountains in its design. The waterfront area itself constantly bubbles with all types of activities, many of them located in the Tom McCall waterfront park, which was reclaimed from a federal highway project. These wonderful projects and activities did not happen randomly; they resulted from the creation of a clear vision of what the Portland region should become.[1]

Hundreds of miles to the east, the first-time visitor to Little Falls, Minnesota, would also discover an area with a strong sense of place, although the community is much smaller in terms of geography and population than the Portland metropolitan region. Walking around the center of Little Falls, the casual observer would notice a growing number of renovation projects in its central business district, such as an historic structure now converted into senior housing or the former railroad depot, designed by nationally renowned architect Cass Gilbert, now renovated for use as the Chamber of Commerce headquarters and community meeting space. Wall murals on downtown buildings depict lifelike scenes from the city's storied past (see Figure 4.1), while decorative lamp standards are being installed in conjunction with road improvements by the state highway department. A beautifully landscaped and maintained public open space called Maple Island Park graces the banks of the Mississippi River near the historic falls. Overlooking the park are two historic turn-of-the-century homes now being used for elderhostels and other community education activities. Like Portland, Oregon, Little Falls is also striving to realize its vision of its future, a vision shaped in part by a Minnesota Design Team visit.[2]

Portland and Little Falls demonstrate that there is still a place for good places. "Now more than ever," write Kotler, Haider, and Rein (1993), "places must think, plan, and act on their futures, lest they be left behind in the new era of place wars" (p. 16). Despite the enormous changes occurring to communities large and small, caused by the rapid pace and enormous scale of the process of post-industrialization, there still remains a place for geographic identity in modern society. In fact, places may even have become more important to disoriented humans in a rapidly changing world. "Protected landscape is a central part of the local economic base," writes economist Thomas Michael Power (1996). "People do care about where they live" (p. 4). The global village is also becoming a globe of villages, and quality of life is more than just a slogan.

Because the special qualities of places like Portland or Little Falls remain significant to individuals, businesses, and communities themselves, communities large and small must now learn to understand and value their unique characteristics, then to chart their own courses within the rolling sea of post-industrial change. Russell Ackoff (1981) of the Wharton School of Business suggests,

> Because of the increasing interconnectedness of individuals, groups, organizations, institutions, and societies brought about by changes in commu-

> nications and transportation, our environments have become larger, more complex, and less predictable—in short, more turbulent. The only kind of equilibrium that can be obtained by a light object in a turbulent environment is dynamic—like that obtained by an airplane flying in a storm." (p. 4)

Like the famous flight of Little Falls' native son Charles Lindbergh, achieving this precarious and ever-shifting balance requires discipline, thorough understanding of both the plane and the forces acting upon it, and a very clear sense of direction.

"Unfortunately, relatively few communities have recognized the critical importance of [community design] in this process," observes planner Randall Arendt et al. (1994), "and have failed to connect their regulations with any overall . . . vision of what they would like to become . . . " (p. 8). Attempting to respond to an ongoing barrage of development proposals, far too many communities allocate scarce local resources for economic development, environmental quality, and social issues in a vacuum, without guidelines envisioning how all of these factors affect one another. They are essentially flying blind.

Communities cannot effectively act on the future while gazing into the rearview mirror of their industrial past. A vision statement is the community's flight plan into the future, its image of how to integrate the many roles and functions of the community into a purposeful system. As economist Kenneth Boulding once noted (1961, p. 6), behavior depends on the image community members hold of a given situation. The visioning process of community design helps to make all its systems work together, to make the flight as smooth and as pleasurable as possible for everyone on board.

Because of the rapid pace of change occurring due to global restructuring, the nature of community planning has also been forced to change. Whereas communities formerly developed a twenty-year comprehensive plan, the time horizon for the future has now been drastically foreshortened. The cybernetic revolution of rapidly flowing knowledge, capital, and ideas means that long-range planning by large units of government is increasingly outmoded. It needs to be supplemented, if not replaced, by smaller, more flexible action teams (Friedmann, 1973, p. 17). These action teams must try to link knowledge and action into a community strategy for shaping the future. Community design, or visioning as it is increasingly called, becomes less a way of preparing comprehensive land use plans and more an innovative method of changing the entire process of community decision making.

Community design is no longer just an activity carried out by professional planners. Drawing once again on systems theory, community design seeks to involve all sectors of the community and build partnerships between government, business, and community institutions. "The holistic principle," says Ackoff (1981), "states that the more parts of a system and levels of it that plan simultaneously and interdependently the better" (p. 74). Based upon this holistic principle, Minnesota Design Teams have included biologists, anthropologists, economists, community and economic developers, and tourism specialists, in addition to architects, landscape architects, and planners, to build a more holistic understanding of community systems.

Community design also assumes that citizens of a community possess the ability and own the responsibility to shape their own future. Places as different as Portland and Little Falls have discovered that envisioning new futures for their communities releases enormous energy and creativity.[3] The community design process functions much like an electrical transformer, converting potential energy or resources available in one form into more productive or desired ones. It offers several advantages not found in the typical planning process in that it

- addresses a wider spectrum of community concerns;
- identifies strategic trends and forces;
- seeks an understanding of basic community values;
- offers a big picture to guide immediate decisions; and
- creates appropriate tools and techniques.

Achieving these crucial objectives requires a variety of tools and techniques. The community design process that has evolved from the case studies conducted by the Minnesota Design Team offers concerned people who want to rebuild or reinvent community through community visioning with a highly useful tool kit for the ultimate home improvement project.

THE MINNESOTA DESIGN TEAM APPROACH TO COMMUNITY DESIGN

A Minnesota Design Team visit is one approach to community design that has been applied in over seventy communities during the past fifteen years. A typical Design Team visit involves months of careful preliminary preparation by team leaders and community members. After being selected for a visit on the basis of its application, the town then provides the team with valuable background data about itself, such as base maps and a community inventory. The next phase involves a three-day design *charrette,* which begins on a Thursday evening when team members arrive in the community. Host families from the local community accommodate design team members at their homes, invariably pampering them and feeding them far too well. Then the real work of community design begins.

The concept of community, the Minnesota Design Team has learned, is complex and constantly changing. Team members, therefore, spend Friday of the charrette weekend listening to and observing the community through a variety of research lenses to develop a full understanding of the life of the community. They learn about the community from multiple perspectives, such as SWOT analysis, focus groups, visual assessment tours, and town meetings. A clearer image of the living community gradually begins to come into focus.

Building on the ancient maxim that breaking bread together is one of the best forms of community building, Friday's activities conclude with a community-wide potluck dinner and town meeting open to all community members. The town meeting, referred to as "democratic brainstorming," uses nominal group process to minimize some typical small town communication barriers. It literally helps bring issues to the table and encourages a full and open discussion of underlying community issues.

Throughout Friday, team members document this wide-ranging exploration of community issues by means of facilitation graphics techniques learned from community consultant Daniel Iacafano.[4] This visual record of the community's concerns functions as a feedback mechanism that helps integrate the issues facing the community while also stimulating new ideas and connections for further consideration. Facilitation graphics also creates a visual archive that community leaders and citizens can draw upon for reference in the future.

On Saturday, the Minnesota Design Team pulls together the background information and community concerns into a visual framework through an intensive charrette. Beginning with a storyboard of an overall concept plan, the team members try to mirror back to the community what they have seen and heard up to that point within a design framework that builds on the unique characteristics of the community. Team members then further develop the key elements of the design framework, such as the town center, waterfronts, open spaces, trail ways, and new development opportunities into a series of graphic images such as maps and drawings.

A Minnesota Design Team presentation at the town meeting on Saturday evening attempts to stimulate widespread community interest in its possible futures by effectively communicating the key design principles and ideas that have emerged during the past few days. When the Saturday evening presentation proves successful, discussions with community members often range far into Sunday morning. Later on Sunday morning, Design Team members typically join their host families for brunch, where additional brainstorming and networking occurs. Team members return to their homes and workplaces, while the community must then decide what to do about its new vision of the future.

Some members of the Minnesota Design Team usually return for a follow-up visit within a year to review the community's activities since the time of the weekend visit. The review session offers an excellent opportunity for team members to remind the community of the key insights and design principles from the weekend, point out current organizational and/or communications problems, and help members of the community identify additional resources or contacts to help them implement their future vision. For example, Little Falls citizens had enthusiastically generated several separate applications to the same funding source for funds to implement projects growing out of the Design Team visit. The follow-up team suggested a regular roundtable session among community members working on Design Team issues and projects to increase communication and build collaborative efforts. Little Falls subsequently became a model for how to implement a community vision.

Like any approach to community design, the Minnesota Design Team possesses both strengths and weaknesses. In particular, the short but intense time span of the charrette can generate unusual levels of citizen involvement but also gloss over underlying problems. Nevertheless, its work with over seventy communities during the past fifteen years has created a considerable body of knowledge about community design, what tools belong in its tool kit, and how to use them most effectively.

ACTION RESEARCH: A KEY TOOL FOR COMMUNITY DESIGN

The renowned researcher William Foote Whyte (1991b) concluded,

> Those aiming to help organizations carry through major processes of sociotechnical change have come to recognize the limitations of the professional expert model. In such situations, we need to develop a process of change, resulting in organizational learning, over a considerable period of time. (p. 9)

The concept of action research effectively addresses Whyte's concerns. It offers a valuable community design technique that can provide communities with some of the vital information and knowledge they need to act successfully on the crucial issues they themselves have identified.

As previously indicated, the Minnesota Design Team requires communities that apply for a visit to describe in their application the major issues they have identified that are affecting their future well-being. In addition, the Design Team conducts an open screening meeting in the community before actually deciding to undertake a design charrette there. In many cases, the Design Team uses this screening visit to forcefully emphasize to community leaders that community visioning must include more community concerns than just economic development or downtown beautification. The workshop helps both the Design Team and the community to more fully understand some of these issues and to encourage interested citizens to share other pressing concerns. These community concerns then become the starting point for studying the community through what is called action research.

Action research involves research about a community that is based on shared community concerns and also involves the community itself in answering the very important questions it poses. It assumes that people learn best when they are engaged in answering important questions they themselves have helped to frame. According to Voth (in Blakely, 1979), action research

> is research used as a tool or technique, an integral part of the community . . . in all aspects of the research process, and has as its objectives the acquisition of valid information, action, and the enhancement of the problem-solving capabilities of the community. (p. 72)

The community, not professional planners or designers, identifies the key issues to be researched and addressed in the community design process.

Community action research can aid considerably in the task of building new understandings of the "messy order" of community. Action research was originally developed in industrial settings by organizational development consultants; it emphasizes involving those being studied in the process of directing change.[5] A major departure from traditional research practices into human society based on detached observation by an observer who formulates the questions, action research tries to encourage heightened awareness and appreciation of the interrelated fields (or environments) of community life by making community concerns themselves the focus of research activities.

In fact, more democratic decision making and participation is one of the basic purposes of action research itself. Ackoff (1981) argues that "professionals should provide whatever motivation, information, knowledge, understanding, wisdom, and imagination are required by others to plan effectively for themselves" (p. 66). Although drawing on the knowledge and skills of technical advisers such as the Minnesota Design Team, action research remains under the control of the community. Its fundamental purposes include both obtaining useful information and improving the decision-making capacities of the local community in the process. It involves citizens in the study of their own community, placing ultimate responsibility for the use of that information on the citizens themselves.

Action research may consist of gathering essential data about the community such as demographic and ecological information, or it may involve obtaining feedback on decision making and the community design process itself. Furthermore, because so much of everyday, experiential knowledge held by the community is not written down and reveals itself mainly through speech and dialogue, finding and using this knowledge also necessitates face-to-face encounters in a number of forums between community designers and the community members themselves such as focus groups, potluck dinners, and town meetings (Friedmann, 1993; Luther, 1990). All citizens can be experts about their own communities, and the action research process draws more people into civic engagement to share their expertise.

"Those aiming to help organizations carry through major processes of socio-technical change," writes Whyte (1991b), "have come to recognize the limitations of the professional expert model" (p. 9). Action research, on the other hand, allows and requires community designers to assume the role of guides rather than experts. The community designer acts more like a coach in building a team approach (Whyte, 1991b, p. 40). Although community designers also participate and contribute their expertise when needed and appropriate, action research operates upward from the citizens, allowing the community to form its own understanding of its situation. Community designers can and do help organize research tasks, pose meaningful questions and research hypotheses about what is happening, challenge and mediate among differing perspectives within the community, and summarize research findings (Friedmann, 1973, p. 184). In doing so, the academic and professional knowledge of the community designers combines synergistically with the more subjective but equally valuable experiential knowledge of community members to create a new understanding, or image, of the community's situation.

TAKING INVENTORY: STUDYING THE SYSTEMS OF COMMUNITY

Envisioning more sustainable futures for communities requires new approaches and a fundamental rethinking and re-searching of the concept of community itself. One key aspect of communities that needs fundamental rethinking involves how to understand their place in the natural world. Urban designer Michael Hough (1995) contends that "the perceptual distinction between city and [nature] has been a root cause of many social and environmental conflicts and the lack of attention to the environment of cities where most problems begin" (p. 1). Communities, it seems, no longer know their place.

As the Minnesota Design Team has discovered over time, many local community concerns about the negative effects of growth spring from the growing separation from nature that Hough thoughtfully describes. Loss of open space and surrounding farmlands, disappearance of familiar natural areas and their replacement with generic subdivisions and commercial strips, along with concerns about water quality associated with the failure of private septic systems drive many of the applications received by the Minnesota Design Team. Like Faust, communities have sought control over nature by dividing land up into individual parcels for easy sale. However, such an approach frequently results in disastrous consequences for both nature and themselves.

Not surprisingly, this alienation from nature also extends to much of the research conducted in this culture by its higher education institutions. This fundamental flaw threatens to make much of higher education irrelevant to many of the pressing concerns identified by many of our communities. Research, in the words of John Friedmann (1973), "has become unhinged from action, leading knowledge to take refuge in the cloistered irrelevancies of esoteric language" (p. 192). Social science in particular has tended to develop explanations that are abstracted from the specifics of time and place, reducing human activities simply to numbers and mathematical models that relate poorly to pressing community concerns.

Although many communities desperately seek useful knowledge and information to guide their decision making, they often find academic studies irrelevant and often unreadable. One key problem facing community decision makers is that typical academic studies based in separate disciplines do not really address the systematic impacts of global restructuring on local communities. Dissecting a frog provides some knowledge of frog physiology, but it's not really a living frog anymore. Although the cybernetic revolution has made us aware of our ecological, social, and economic relationships, "studies of single impacts," write Godschalk and Brower (1989), "do not reflect the full range [and interrelatedness] of growth management [issues]" (p. 160). At some point, the frog has to be re-membered and released into the pond again.

Pioneering planners and systems thinkers such as Sir Patrick Geddes and Robert Park strongly maintained that the design of cities and communities required what Geddes termed a *synoptic,* or synthetic, view. Geddes, for example, emphasized (1973) that life had to be lived in the real world, that interconnections had to be made between the world of learning and the world of living. In his advocacy of *transactive planning,* planning theorist John Friedmann (1973) echoes the concerns of Geddes and Park for working within a holistic framework. "The problems on which [community designers] work," writes Friedmann, "must be studied in the fullness of historical circumstances. The number of variables that must be considered is substantially greater than those included in the analytical models of [social] scientific work" (p. 183). Communities must always be recognized as living systems.

As we have previously seen, the pond for most communities today involves the sea change of post-industrialization. The impacts of global restructuring upon local development appear in many different yet interrelated areas: demographics, housing activity, land use, environmental issues, social organizations, and many others.[6] During a follow-up visit by the Design Team, the leader of the Paynesville, Minnesota, Chamber of Commerce mentioned that a large company had located in Paynesville primarily because it was quite impressed by the downtown beautification activities the community had undertaken. The key question for community design, then, becomes how to understand communities as complex social systems and to gather detailed information about key aspects of them while still recognizing and reinforcing the interrelatedness of the physical, economic, and social elements of these living systems.

There are means to accomplish this task. According to organizational development expert Warren Bennis (1993), "the way in which organizations can master their dilemmas and solve their problems is by developing a spirit of inquiry" (p. 57). This spirit of inquiry needs to begin with community environmental concerns. The growing realization that local communities depend on their natural as well as social environments makes it increasingly clear that current patterns of development are fundamentally mistaken and cannot indefinitely continue. George Brewster (1996) of the Urban Land Institute, a professional organization that deals with the concerns of the real estate development community, concludes that "with the realization that the world, and the environment, are part of a larger whole on which we all depend, it is apparent that our patterns of development are based on an outmoded model of reality. This new [ecological] model makes us aware of our ecological, economic, and social interconnectedness" (p. 25). Community environmental concerns now require researchers to open their conceptual boundaries and include the living, organic environment itself as part of the research problem. As Robert Park observed, the city is part of nature, especially human nature. We are not outside of the environments that we study.

This new understanding of community environments demands a comprehensive, systems approach to the issues affecting a given community. Ecologist Aldo Leopold remarked (in Bradley, 1996) on the woeful lack of communication between naturalists studying biological communities and social scientists studying the human community. "The inevitable fusion of these two lines of thought will, perhaps, constitute the outstanding advance of the twentieth century" (p. C19). Both lines of thought are essential to effective action research.

The guidelines established for the National Environmental Policy Act represented a groundbreaking attempt at creating such a fusion of the natural and social sciences, as well as a valuable approach to rethinking community design. According to the National Environmental Policy Act (NEPA) of 1969 (P.L. 91-190), all agencies of the federal government are required to use a systematic, interdisciplinary approach to planning and decision making for major federal projects (Ditton, 1973; Mandelker, 1984). This approach was expected, as Leopold hoped, to integrate the use of natural and social sciences as well as the design arts into a systematic approach. Although seldom applied in such a holistic manner, NEPA points the way to an appropriate methodology for how community design should be carried out (Bass, 1993; Clark & Herington, 1988). It also offers an excellent vehicle for bringing professional research knowledge to bear on pressing community concerns in the form of action research.

Like Aldo Leopold and NEPA, the Minnesota Design Team also seeks to integrate the natural sciences, social research, and the design arts in attempting to help communities understand themselves and guide their future decision making. A detailed community profile or inventory provides the necessary foundation to achieve such an integrated vision. To facilitate the process of conducting such a profile or inventory, the Minnesota Design Team developed a series of action research questions based on the NEPA guidelines (see Figure 4.1).

This series of questions is based on an understanding of a community as a living organism made up of several related environments (or systems), starting with and building on the natural environment itself. Ackoff (1981) suggests that the "proper role of a social system is to encourage and facilitate the development of its members. Doing so requires that it carries out four functions: the scientific, the economic, the ethical-moral, and the aesthetic" (p. 49). These categories translate fairly readily into the natural, economic, social, and cultural environments of a community. As part of their preparations for the Design Team visit, communities are expected to form action teams to investigate and report on these four vital community systems. Action teams offer an excellent opportunity to build a detailed knowledge base of key aspects of community, showcase the diversity of talents and expertise that already exist within a community, and involve citizens who might not otherwise participate in civic projects. Asking volunteers for a short-term time commitment to research an aspect of the community they care about, such as wildlife habitat or educational programs, also offers an ideal vehicle for community building.

The results of such an inventory can be quite impressive and their impacts long-lasting. The town of Caledonia in southeastern Minnesota assembled a thoroughly researched notebook for the design charrette

The purpose of these questions is to: (1) provide an environmental assessment needed for effective community design; (2) help your community develop its capacity to understand itself more fully and to act upon that understanding.

A. The Natural Environment:
1. What *geological forces* have shaped your region?
2. What *hydrological forces* have shaped your region?
3. Characterize the *topography* of your region.
4. What characterizes the *wildlife* of your area?
5. Identify and describe local *wetland areas.*
6. How would you characterize your *climate*?
7. Identify and describe the main types of *vegetation* in your region.
8. Identify and describe the major *pollution problems* in your area.
9. What are the *major land uses* in your community?
10. Identify and describe *major local environmental organizations.*

B. The Social Environment:
1. How has your *population* changed over the past three censuses?
2. How many *males and females* are in each age bracket?
3. What is the *household composition* of your community?
4. What are the main *ethnic and racial groups* in your community?
5. What is the *education level* in your community?
6. Describe the *income levels* in your community.
7. What are the *major religious denominations* in your community?
8. Characterize the *housing stock* of your community.
9. What *types of housing* are inadequate?
10. Identify and describe the *major community service organizations.*

C. The Economic Environment:
1. What is the *trade area* of your community?
2. What is the *industry mix* of your community?
3. Who are the *major local employers* in your community?
4. Identify and describe the *major transportation facilities* in your region.
5. Identify and describe the *major public utilities* in your community.
6. What is the *employment rate* in your community?
7. Summarize your *most recent shopper survey.*
8. Describe the *labor force* in your community.
9. What *business incentives* does your community offer?
10. Identify and describe *local economic development organizations.*

D. The Cultural Environment:
1. Describe the *key events* in your community's history.
2. What *key individuals* have shaped your community?
3. Identify and describe interesting *local customs and traditions.*
4. Describe popular community *folklore and legends.*
5. What are the *major community festivals*?
6. Describe any *local historic districts.*
7. Identify and describe *local historic buildings.*
8. Identify the *best views* of and from your community.
9. Identify the *special places* of your community.
10. Identify and describe *local cultural organizations.*

Figure 4.1. Minnesota Design Team Community Action Research Guidelines

about each of its community systems, including demographic data, economic and social information, historic sites, and detailed information about the natural history of the region. A diverse group of citizens also made presentations to the Design Team to explain their findings and answer further questions. Such systematic research not only provides community designers with an extremely valuable database, but it also creates a corps of knowledgeable community residents who can answer the community's future questions about itself.

FORCE FIELD ANALYSIS: TAKING ANOTHER SWOT AT THE COMMUNITY

Force field analysis is another key element of action research that assists a community with taking inventory and enlarging its understanding of itself. It is closely related to the concept of strategic planning regularly used in the business world. Whereas community visioning starts with a blank slate, so to speak, allowing citizens to move beyond the confines of existing images of their community, strategic planning begins from the premise that no actions occur in a vacuum. Consequently, this element of community design emphasizes thorough analysis of the present state or conditions that relate an organization or community to its surrounding contexts, to help it find or create its most suitable niche in the dynamic post-industrial future. Such analysis focuses upon both internal factors, such as how the organization or community operates, and external forces acting on the community, such as who holds a stake in certain decisions or actions.

Force field analysis owes considerably to the work of two influential twentieth-century social scientists: Karl Mannheim and Kurt Lewin. The work of sociologist Karl Mannheim (1945) on the need for clearer understanding of complex social systems provides a valuable foundation for understanding the concept of force field analysis. The ability to foresee patterns and forces, according to Mannheim, was crucial to societal survival. He believed that organizations such as communities needed to probe the currents of social change to uncover what he termed the *principia media,* or developmental processes, causing structural changes to the organization. This type of knowledge consisted of intelligence about the operations of complex social systems, their multiple meanings and interrelationships, and their underlying forces and directions.

Force field analysis also borrows heavily from the work of anthropologist Kurt Lewin about the social psychology concept of force fields. "The field theory of Kurt Lewin," writes Kenneth Boulding (1961), "is also a clear example of the image in social-psychological theories" (p. 151).

Lewin (1951) believed that organizations operate within social psychological environments called *fields*, which effectively serve as images, or boundaries, for what groups will do or accept. Examples of fields within a community include its demographics, organizations, values and traditions, and existing resources; these fields can either reinforce one another or conflict among themselves. Because these various forces, or fields, in a community can substantially affect the process of change, the need clearly exists to increase understanding of the force fields that are operating on the organization or community (Hustedde & Score, 1995).

The SWOT technique frequently employed in strategic planning functions effectively as action research into the force fields acting within or on a given community. SWOT stands for:

- Strengths
- Weaknesses
- Opportunities
- Threats

SWOT analysis represents a systematic attempt on the part of a community to self-critically examine how it actually operates (Fischer, 1989; Schoemaker, 1995; see Jones, 1990, for a valuable application of this process to urban neighborhoods). It may be the most difficult yet most valuable part of the community design process. Stepping outside one's environment is always a difficult maneuver to execute.

The Minnesota Design Team began employing SWOT analysis several years ago in an effort to increase its useful knowledge about what was actually happening in host communities. Prior to conducting SWOT analysis, the Design Team would typically spend its Friday mornings listening to civic leaders talk about their communities. In a three-day design charrette, time becomes a highly critical factor; the Friday morning sessions seemed especially unproductive in terms of moving the community design process along. Many of the presentations often fell into the category of civic boosterism, explaining how great the community was and painting a glowing picture of its future. Although civic pride represents one of the goals of community design, such boosterism typically avoided addressing underlying problems that had led the community to invite the Design Team in the first place.

The Minnesota Design Team finally required communities to have representatives from a wide range of community groups identify and speak about what they considered to be the major strength, weakness, opportunity, and threat facing the community from their perspectives. Freed from the force field of community pressure to offer nothing but praise, speakers were then enabled to provide a more balanced assessment of their communities.

In addition, recording these assessments in a SWOT matrix format began to reveal patterns or forces acting within or upon the community. Revealing these patterns and forces has helped the Minnesota Design Team to more quickly and clearly understand the primary force fields within a community. For example, SWOT analysis in Clearwater, Minnesota, disclosed that many people in the community clearly wanted to reestablish connections with the town's historic river front along the Mississippi River. This insight subsequently guided the Design Team's field survey work, its questions at the Friday evening town meeting, and its subsequent design recommendations.

BENEFITS DERIVED FROM ACTION RESEARCH

Action research offers a number of important benefits for both community designers and community members that differ from traditional planning practice. A number of examples from Minnesota Design Team communities illustrate these benefits. First of all, community design allows the community to explore and use its own value system. Ian McHarg (1971) writes,

> [Community design] permits a most important improvement in planning method—that is, that the community can employ its own value system. Those areas, places, buildings or spaces that it cherishes can be so identified and incorporated into the value system of the method. (p. 104)

For example Lake Elmo, Minnesota, identified the surrounding farms and open spaces as key elements of the community to be preserved and protected, whereas Little Falls emphasized preserving its historic Main Street and scenic views of the Mississippi River.

Second, community action research provides intelligence, data combined with insights that would not ordinarily be available to outside planning experts. In the process of conducting action research, all community members become experts. Citizens in several Design Team communities who ordinarily might remain in the background have emerged to share important and hidden knowledge about the natural, social, and cultural environments of their communities. For example, local anglers and hunters often reveal important information about loss of habitat and growing pollution concerns. Action research results in much more detailed knowledge of the community than would otherwise be available.

Third, community involvement in action research creates a sense of shared ownership in the process of community design. For example,

Figure 4.2 Community Volunteers Created a Magnificent Bandstand in Caledonia

citizens in the rapidly growing town of Becker uncovered the fascinating history of their Main Street in the process of carrying out community action research. Intrigued by the role of the railroad in shaping Main Street, citizens formed a "T-Town" committee to follow up on their findings and to work to preserve some of its historic structures from encroaching development. In Caledonia, participants discovered the importance of community building to the early Scottish founders of the town. They subsequently began to emphasize the need for the community to tell stories about its origins and to rekindle the spirit of the town's founders. A community volunteer effort subsequently created a magnificent bandstand in the local park (see Figure 4.2).

Finally, community action research also increases the community's capacity to carry out subsequent actions based on its findings and work with external agencies. For example, the community of Taylors Falls conducted a considerable amount of background research about issues involving the creation of a pedestrian underpass that would link their Main Street with a state park nearby. In the process, they worked with state officials at great length and gradually developed an effective working relationship with the Minnesota Department of Transportation. This working relationship helped Taylors Falls to successfully argue their case for the underpass project. They then collaborated with the Minnesota Department of Transportation on a major design project that won an award from the Minnesota Chapter of the American Society of Landscape Architects.

In the community design process, the knowledge of both citizen and designer undergoes significant transformations. A common image of the situation begins to evolve through mutual dialogue, and people begin to envision new possibilities for action. Empowered with this new knowledge, people better understand the forces acting on their communities and become more willing themselves to act on their research findings. They begin to create community.

NOTES

1. I am indebted to City of Portland city planner Mark Bello, Ph.D., AICP, for his thoughtful insights into the Portland community design and visioning process, offered at the May 3-5, 1995, Urban Affairs Association annual conference held in Portland, Oregon.

2. I am indebted to Ms. Cathy VanRisseghem, director of the Little Falls Convention & Visitors Bureau and a member of the Design Team in Little Falls, for her insights into the role of community visioning in the redevelopment of her community.

3. Oregon has been at the forefront in the use of community visioning. It has developed a four-step model of community called, not surprisingly, the Oregon Model (see Ames, 1993).

4. The Minnesota Design Team owes an enormous debt to Daniel Iacafano. He led a workshop for the Design Team in the fall of 1990 that proved to be a milestone in its understanding of community design and considerably strengthened its technical capabilities.

5. The works of Chris Argyris and Warren Bennis are especially valuable in understanding the concept of action research. See especially Argyris (1985) and Bennis (1993).

6. While working with a group of local economic development specialists, I was pleasantly surprised to hear one of them say that he couldn't separate economic development from community development any more.

CHAPTER 5

COMMUNITY IN THE THIRD DIMENSION

One of the major weaknesses of traditional city planning and academic research approaches to the study of communities involves the failure to recognize the vital importance of a community's physical appearance to both its residents and visitors. "Professional planners, with their urgent need to act, move too quickly to models and inventories" (Tuan, 1977, p. 7) and often "tend to screen out the connections between the physical environment and its social meaning" (Appleyard, 1979, p. 143). Far too many comprehensive plans or economic development strategies focus their efforts entirely on issues of land use or public finance, virtually ignoring environmental and visual considerations because they are far less easily quantified or considered simply matters of taste.

Communities, however, do not just exist as numbers and statistics, regardless of how valuable the insights they offer to a keen observer. They also exist in the third and fourth dimensions, involving all our senses including the sense of passing time (Lynch, 1982). A holistic, or systems, approach to community design also needs to come to terms with what architectural historian Christian Norberg-Schulz (1979) calls the *genius loci,* or characteristic spirit of a place, in order to be truly successful.

A significant and growing body of research in the field of environmental perception and design clearly demonstrates that communities do indeed possess such a genius loci. In addition to their functions within natural, economic, and social contexts, communities also exist as three-dimensional, spatial environments that their citizens perceive daily as either sources of delight, displeasure, or despair (see, for example, Rapaport, 1982; Sinha, 1995; Walther, 1988). "Over the last 25 years," writes environmental psychologist Linda Groat (1995), "many architects and other design professionals have come to recognize the importance of the various meanings people associate with the physical environment" (p. 1). Thinking fully and carefully about the visual aspects of a community and raising critical questions about them is an extremely important element of successful community design.

KEY ISSUES OF COMMUNITY APPEARANCE

In addition to its impacts on the socioeconomic environments of communities, post-industrialization has also contributed to the deterioration of the visual environment. "We are threatened today by two kinds of environmental degradation: one is pollution—the other is loss of meaning. For the first time in human history, people are systematically building meaningless places" (Walther, 1988, p. 2). Edward Relph (1976) coined the provocative term *placelessness* to describe this type of modern urban development that lacks any distinctive character or relationship to its site.

Even in a post-industrial society and global marketplace, however, place is still important to many people. The concern about the loss of meaningful places is much more than an academic issue. In the words of cultural geographer Yi-Fu Tuan (1977), "space and place are basic components of the lived world; we take them for granted. When we think about them, however, they may assume unexpected meanings and raise questions we have not thought to ask" (p. 3). Many local residents bemoan the loss of small-town atmosphere or neighborhood in their community, the sense of a distinctive physical and cultural setting that differs meaningfully from the development patterns associated with large metropolitan centers. What has happened to cause this loss of meaning in so many special places?

The systems approach to the study of community ecology also applies to analysis of the physical environments. It suggests that major changes in some aspects of the overall community system, such as replacing traditional forms of industrial jobs with service activities or introducing new forms of residential patterns, necessarily show up at some point in the appearance of the community's built and natural environments. All elements of a community system are interconnected. Many Minnesota Design Team communities now find themselves wrestling with the difficult transition from farming and industrial jobs to retailing or tourism employment. They also are attempting to come to terms with proposed new subdivisions that differ dramatically from the old grid pattern of the original town plat, and with burgeoning fast-food franchises and discount stores being located on the edge of town. Not surprisingly, then, for many community members the appearance of the built environment becomes the focal point for many of their community design activities and the main stimulus for community change. Nasar (1990) observed that although "visual quality alone may not justify environmental change . . . it should not, however, be overlooked on the grounds that it is a minor concern, a matter of taste. [Research] suggests that the public cares about and agrees on city appearance" (p. 50).

Experience with more than seventy communities over the past fifteen years has given the Minnesota Design Team an extended opportunity to learn firsthand about the issues facing small communities as they wrestle with the impacts of rapid economic and social changes. As part of the application process for a Design Team visit, communities are asked to respond to open-ended questions such as "List and briefly describe the three most important problems for your community at this time." Although their responses vary widely because of their unique geographic settings and social characteristics, concerns about the visual appearance of their communities almost invariably receive prominent mention in their applications. As Nasar's research suggests, Minnesota Design Team communities do indeed care about and generally agree on key issues of community appearance. Although beautification and a general sense of order and attractiveness represent overall concerns, the issues concerning community visual appearances tend to group themselves into four basic categories:

1. *Community gateways* generate high levels of citizen concern because they provide some of the most visible and important visual clues to the character of a community. For many visitors, the "edge of town" in a community creates that critical first impression, which they often find very difficult to ignore or go beyond. "The external image a city presents to the world," says urban designer Richard Hedman (1984), "is the signature by which it is known" (p. 105), and planner Randall Arendt and colleagues (1994) contend that "public perception of community character is based largely on what can be seen from the automobile" (p. 192).

Many Minnesota Design Team communities have expressed major concerns in their applications about the appearance of their gateways and the potentially negative images those gateways present to travelers and visitors. Ironically, applicants seldom complain about the effects that unattractive entrances have upon residents themselves, although these effects may be even more important than those on visitors. These community edges are often characterized by fast-food franchises and filling stations, the standard array of national franchises found all across the American countryside. Totally oriented toward the automobile and posing major health hazards to the few unlucky remaining pedestrians, this type of development devours local farms and open spaces, filling the area with the inducements of the commercial strip. Local place identity disappears amid a tangled maze of neon signs and advertisements that could be found anywhere throughout the country and increasingly in other parts of the world.

2. Meanwhile, the *waterfronts* that historically gave birth to local communities have receded far into the background of their awareness until flooding or serious water quality problems literally bring them to the surface once again. Long abandoned as the town's main commercial district, these areas have gradually become dumping grounds for much of the unwanted debris generated by the community. They may have become sites of dumps, landfills, rundown housing, or warehouses and are frequently fearsome places where few local residents dare venture. Some people from the community remember the time when the water was clean enough for fishing and the area still pulsed with business and other elements of community life; many others wonder what the town's waterfront could once again become in the future.

3. The historic *Main Streets* of these communities typically generate much of the impetus for efforts at community design. These once-important places and social centers now often appear sadly run-down and sometimes nearly abandoned. As the model of community ecology suggests, uncertainty about the function and the future of the old Main Street contributes further to its poor visual appearance. Storefronts lack fresh paint or other visible forms of routine maintenance, old signs touting obsolete products slowly fade a bit more each year, and for-sale or lease signs sprout like mushrooms in the decaying atmosphere. Crumbling historic structures offer only sad testimony to a once-proud past. New businesses moving to the community typically seek the highly visible locations at the edge of town near the main highway interchanges and industrial park, where easy access to automobiles and large tracts of available land made available by the city encourage new development to occur. The Main Street district fades further into oblivion.[1]

4. The appearance of older *residential neighborhoods* or occasionally mobile home parks provides another visual clue to the condition and trends of the applicant community. As new residential construction occurs near the edges of town on former farmland, older and historic residential areas do not receive much-needed infusions of people, resources, and attention. Like the historic waterfronts and Main Streets,

these once-special places continue their slow process of deterioration. They may exist in the minds of local residents as problem areas, evidence that something has gone seriously wrong in the life of the community, or they may be completely out of sight and mind. Ignored and suffering from a major lack of reinvestment, these older residential sections cannot effectively provide affordable housing for first-time home buyers, thereby adding to the serious housing shortages facing many small towns and urban neighborhoods.

RELEVANT ACADEMIC RESEARCH ON VISUAL IMAGE

The built environment and its associated meanings have generated considerable research interest. "It was architects and building engineers who first approached psychologists for advice on the impact that their designs might have on their users" (Canter, 1995, p. vii). Over time, however, more and more research has emerged to help understand how ordinary citizens perceive their communities and create their own perceptual worlds. A considerable body of academic research has subsequently been created to address these vital concerns about the built and natural environments. The research clearly demonstrates the nature and vital importance of such key design elements like Main Street and waterfronts to a community's understanding of itself as well as its overall sense of well-being. New techniques based upon this exciting field of research offer community design practitioners important new windows into the lives and psyches of local communities.

The Experience of Place

The first important category in this body of research deals with the complex ways in which human beings experience and make sense of their spatial environments. One of the leaders in the field of cognitive research was the late Kevin Lynch. Lynch (1960) maintained,

> Although clarity or legibility is by no means the only important property of a beautiful city, it is of special importance when considering environments at the urban scale of size, time, and complexity. To understand this, we must consider not just the city as a thing in itself, but the city being perceived by its inhabitants. (p. 3)

His pioneering studies on the imageability of communities demonstrated the importance of creating a clear framework of key design elements that would enable residents (and visitors) to visually understand and "read" their community. Lynch maintained that people feel most comfortable in an environment when they can clearly perceive its underlying order and patterns. According to Lynch, the principal benefits of a clear visual image for one's community include the following:

- Ease of mobility
- Broadened frame of reference
- Community symbols and collective memories
- Sense of emotional security
- Heightened sense of human experience

However, Lynch's (1960) concept of imageability did not specifically address the critical question of the symbolic meanings that people assign to these design elements. "These will be glossed over," he wrote, "since the objective here is to uncover the role of form itself" (p. 46). Although Lynch readily acknowledged the critical importance of this issue, he felt that it was beyond the scope of his research at the time of his writing. Nevertheless, the meanings people assign to their physical environments may actually hold the key to creating a satisfying community image.

Cultural geographer Yi-Fu Tuan (1977) thoughtfully explores the phenomenon of how spatial meanings are created in *Space and Place: The Perspective of Experience.* Tuan draws upon human experience across cultures, ancient and modern, western and oriental, with literate and oral traditions, to examine the human capacity for symbolization, "how the human person, who is animal, fantasist, and computer combined, experiences and understands the [spatial] world" (p. 5). Tuan considers basic human experiences across cultures such as the relationship of the human body to space, attachments to homeland, and the relationship of time and place. His work offers a valuable complement to traditional land use planning approaches, drawing "attention to questions that humanists have posed with regard to space and place" (p. 7).

Landscape architect Randy Hester has further extended Tuan's concepts to include what Hester terms the "sacred places" of a community. According to Hester (1995), every community values certain places above all others. These places are not necessarily the sites of religious worship or even the most attractive ones in the community.[2] They may often be ordinary places that have acquired uncommon significance in the life of the community over time and that the community wishes to preserve and see revitalized. The '50s-style drive-in restaurant in Taylors Falls, Minnesota, where waitresses wearing poodle skirts and gliding on roller skates take customer orders, represents one delightful example of such a sacred space that has acquired deep community associations over several generations. Hester argues that it is especially important to identify and document these special places, even if they are not major design elements of the community, to gain deeper insights into the community's character and to strengthen the sense of community identity (pp. 7-8).

The Patterns of Place

The second key element of a community's sense of place involves the actual physical patterns found in a given community. According to Lynch (1960), a public image of any given city or community exists that is built up gradually through experience. Although each person creates his or her own unique cognitive (mental) map or picture of the community as the cumulative result of personal experiences there, it is the general public image of a city, town, or neighborhood that is "more or less compelling, more or less embracing" (p. 46) and offers the greatest potential leverage for community design. The image of every built environment, Lynch maintained, consists of five key elements: districts, nodes, landmarks, edges, and paths. Each community selects the materials for these elements and arranges them in its own way, like an artist exploring his or her palette of colors, materials, and textures. How these elements are ultimately composed gives each community its special character, or image:

Districts are medium to large sections of a community with a special, recognizable character of their own that people enter; Main Streets and waterfronts, for example, would typically be considered districts.

Nodes are key places and strategic centers within the districts of the community, focal points of much of its activity, such as City Hall or the old post office on Main Street.

Landmarks are usually physical objects that provide external reference points, such as the cupola of a historic church or an old clock on Main Street.

Edges are linear boundaries or breaks in the appearance of the community; the historic waterfront typically represents such an edge.

People move between these elements along the *paths* of the community, whether they are highways, Main Street, county roads, or residential streets. These paths are especially crucial to the formation of the overall image of the community because of the high visibility they offer.

In his influential book, *A Pattern Language* (Alexander, Ishikawa, Silverstein, with Jacobson, Fiksdahl-King, & Angel, 1977), urban designer Christopher Alexander proposed some 250 elements that he and his colleagues think should be considered in any community design process. These elements, or patterns, can be combined in an infinite number of combinations to reflect and create the unique character of each community, but, according to Alexander, they represent basic building blocks that seem to characterize successful communities. Some of these patterns include

1. *Integrating homes and workplaces.* "The artificial separation of home and work creates intolerable rifts in people's inner lives. . . . Concentration and segregation of work leads to dead neighborhoods" (p. 52).
2. *Creating public spaces.* "Somewhere in the community [create] at least one big place where a few hundred people can gather, with beer and wine, music, and perhaps a half-dozen activities, so that people are continuously crisscrossing from one to another" (p. 446).

Unlike the large-scale patterns that Lynch focuses on in his work, Alexander focuses his design lens on the smaller physical patterns within a community that he believes contribute to a more satisfactory sense of place and heightened sense of community identity.

To actually improve visual appearance, however, decision makers need to know how community residents actually feel about the environment and its component elements. People constantly evaluate their environments (Nasar, 1990). Furthermore, in a dynamic community system imageability and how people feel about those community places relate to one another in a continuous feedback loop. People tend to recall those places about which they hold strong feelings, and vice versa. It makes little sense to change visual appearances unless those changes directly affect residents' images of the city, especially how they feel about its key elements.

Certain evaluation patterns concerning how people feel about visual appearances of a built environment have gradually emerged. Nasar's research indicates that people typically dislike chaotic commercial strip development, roadside signs and billboards, dirtiness, run-down buildings, lots filled with weeds and debris, and lots of poles and wires. On the other hand, people generally prefer landscaping, open countryside, scenery and broad vistas, places of historic significance, and a general sense of some design order. Other research efforts support Nasar's findings. Anton Nelessen (1995), one of the leaders in the use of visual assessment techniques for community design, concluded after working with hundreds of communities that most Americans reject the current pattern and spatial characteristics of urban sprawl in favor of more traditional or neo-traditional small communities.

The Politics of Place

The final key category of research dealing with a sense of place involves the issue of who wins and who loses, whose meanings receive community attention and resources. The sprawling commercial strip on the edge of the town can possess very different meanings for different people or social groups such as the local Chamber of Commerce or members of the town's preservation society. Imageability, then, is not just an aesthetic quality or a means of finding one's way around town; it is also the physical expression of the prevailing power and tastes in the community. In short, the visual appearance of the community itself functions as a communications system that explains to observers what the community values.

For designer Donald Appleyard (1979), the concept of social symbolism offers one important way to deepen our understanding of the link between the physical environment and the sociopolitical structure of a community. An environment becomes a social symbol "when it is intended or perceived as a representative of someone or some social group; when social meaning plays an influential role in relation to its other functions" (p. 144). Appleyard presents a communications model of an environmental system in which a particular physical environment or action is depicted as one element of an overall system. The place or action does not exist in isolation but as part of a complex, evolving network that includes the intended messages sent by owners, designers, and managers and the messages received by consumers such as users, neighbors, and the general public.

Appleyard suggests several policy implications for community design that grow out of the communications model (p. 152). These include:

1. The human need for identity and power finds expression in the physical environment.
2. Community design can threaten the identity and status of some groups while enlarging the power of others.
3. Citizen participation in community design allows more people to give meaning to new environmental actions in their places.

As Appleyard indicates, the appearance of the built and natural environments speaks loudly and visibly about who controls decision making in a given community. The work of landscape architect Grady Clay (1980) offers other valuable clues into both local decision making and the macrolevel forces that affect the built environment of a community. In Clay's view, "Topography is so often a clue to social geography" (p. 145). For example, Clay identifies the special tourist districts that have cropped up in many communities, districts that have been created to attract new industry and hordes of free-spending tourists, as examples of myth making and local political control by elites. At a larger scale, the highway commercial strip with its array of fast-food franchises and megastores suggests that power and decision making now lie outside the local community; Wal-Mart has triumphed over Main Street.

Because of the close relationship that exists between visual appearances and community decision making, citizen participation in visual assessment becomes critically important to the success of the community design process. Like action research, it enables residents to better understand and identify with their environments, reduces the growing sense of alienation associated with the process of post-industrialization, and helps residents begin to take responsibility for shaping the community in their own image.

VISUAL ASSESSMENT TECHNIQUES

This growing body of academic literature on environmental design has provided the foundation needed for some innovative visual assessment techniques that can assist in reshaping a community's visual identity. Visual assessments offer valuable insights into the character of a given community, insights that can greatly enrich and improve the community design process. Wade Vitalis, the president of the Taylors Falls Chamber of Commerce at the time of the Minnesota Design Team visit to that community, remarked about the importance of community appearance and visualization to successful community design. Vitalis (quoted in Mehrhoff, 1995) observed that "our community needed someone to draw the pictures so that people could get an image in their minds of what we could look like. We couldn't get past that first step of imagining how things might change" (p. 9). Visual images, therefore, represent an absolutely essential research component of the community design process.

Visual assessments can take a wide variety of forms in the community design process. Some of these techniques are very highly structured, whereas others tend to employ a more open-ended approach that allows the community itself to assign more of the meanings to the visual data.

Evaluative maps are one of the most basic forms for visual assessment of a community. Respondents are asked to locate on a map the physical features that they consider important in the overall scheme of the community, either positive or negative. Gradually, a composite image of distinctive community features emerges, showing how people actually perceive the visual elements of their community. According to Nasar (1990), evaluative maps can suggest the effects of city structure and typical or unusual experiences. Five desirable features clearly emerge:

1. Naturalness
2. Upkeep
3. Openness
4. Order
5. Historical significance

"By showing the identity, location, and likabil
Nasar concludes, "evaluative maps provide a
(p. 41).

The Minnesota Design Team successfully em
ment technique in Becker, Minnesota. Respo
student intern to identify a number of areas v
found the most visually pleasing and the s
found equally unpleasant. Residents were a
general boundaries of these areas as well as th
that led to their evaluations of the areas as eit
Each respondent's evaluative images were re
intern, who then overlaid the maps on one anothe
map showing the overall evaluative image of the resp
resulting composite map helped indicate to Design Team members those
areas of the community that seemed especially significant for further consideration during the actual visit. For example, the historic Main Street (eventually dubbed the "T-Town"), which had developed decades earlier across from the railroad station, emerged as a sacred space that warranted deeper consideration of its place in the overall development of the community.

Visual surveys are another visual assessment technique that possesses considerable value for community design research purposes. This particular technique has been considerably refined by the firm of Anton Nelessen Associates, Inc. A visual survey involves asking community residents to rate paired images of their community as either acceptable or unacceptable. Images that illustrate the real-world characteristics of existing community zoning and design are selected by community designers for purposes of comparison. By reviewing a large number of such paired responses, a visual survey helps to reveal strong preferences and the presence of a general community consensus regarding visual appearances.

The community of Lake Elmo, Minnesota, employed a variation of the visual survey technique while conducting background research for a Minnesota Design Team visit (see Figure 5.1).

Rather than begin with contrasting images selected by designer professionals, Lake Elmo asked its citizens to take two photos, one positive and one negative, in six basic categories such as waterfronts, roads, and special districts. The photographs were then gathered and mounted on foam core board for both Design Team members and Lake Elmo residents to view and assess. Such an open-ended format does not yield the definitive results of the Visual Preference Survey® used by Anton Nelessen Associates, but it does allow communities considerable scope in creating their own value system of places.

ays to understand the opportunities, problems, and beauty of a community. One
people and events that contribute to the sense of community. One can also consider
cology and the surrounding context. Finally, a "sense of place" can be expressed
architecture, landscape, and open spaces of the community itself.

survey, we would like you to photograph places that you feel contribute to the quality of the
Elmo experience. These photo subjects can also illustrate problems that detract from this experi-
ce. The information gained from the survey will be used to help the Minnesota Design Team prepare for the Lake Elmo visit.

Instructions:

Please take two photos of each of the following subjects:

- the lake itself
- the town center
- a residential area
- a natural area other than the lake
- development along a roadway
- what makes the City of Lake Elmo unique

One photo should show a *positive* example, the other a *negative* example.

Use the accompanying forms to describe your reasons for taking each photo.

Return your camera (we will develop the film) and written forms to Ann Terwedo, Lake Elmo City Hall.

All cameras and forms should be returned to Ann by August 30.

Questions? Please call Arthur Mehrhoff, Saint Cloud State University, @ (612) 255-3107.

Minnesota Design Team Lake Elmo Fall 1995

Figure 5.1. The City of Lake Elmo Visual Assessment Survey

Citizen photography projects offer yet another visual assessment technique, combining several elements of the previous techniques. Like evaluative maps, they are open-ended; respondents select the images of the community themselves. However, like visual preference surveys, they require photographs of actual community places rather than generalized descriptions. In many respects, the citizen photography project resembles community action research. Like community action research, the citizen photography project allows community members themselves to create the categories and gather the data necessary to answer their own questions.

For example, citizens of St. Joseph, Minnesota, were given inexpensive panoramic cameras and asked to photograph the places, people, and events that most strongly contributed to their sense of community. They also were asked to photograph places that they felt weakened their sense of community and represented negative visual experiences. The contents of the photographs were then analyzed by the Minnesota Design Team leaders for the St. Joseph visit. A number of categories clearly emerged from the analysis of the photographs:

- Valued places
- Growth patterns
- Town entrances
- Design opportunities

Such a summary, however, requires considerable time and lacks a high degree of statistical reliability. Analysis and interpretation of the community photographs demand highly discriminating judgments on the part of the reviewers; categories can sometimes blur in the blizzard of images. Like action research, however, the citizen photography project can provide both useful data about the community and increasing citizen involvement in and ownership of the community design process.

For example, the St. Joseph citizen photography project generated valuable evaluative images of the community and a strong community consensus about key design issues. In addition, the photographs were subsequently mounted on display boards and the series exhibited both during the Minnesota Design Team visit and in various businesses and civic buildings around the community following the visit. The photographs appear to have taken on a life of their own, acquiring and adding new meanings to the process of community design as citizens continue to discuss and add to their importance.

The Minnesota Design Team also conducts its own visual assessments both before and during the weekend design charrette to act as a kind of cross-reference to the community visual assessments. This visual reconnaissance allows community designers to compare their images of the community with the community's own, both in terms of shared assessments and at key points where team members' perceptions differ significantly from those of the community. The reconnaissance tour also provides the community with a view of how a visitor might initially perceive it.

The visual reconnaisance actually begins when team leaders first visit the community to discuss the Minnesota Design Team visit and to further explore the issues facing the community. In addition to the photographs submitted by the community as part of their application for a visit, team leaders typically take pictures of striking physical features or glaring problem areas that they observe. These images are usually shared with members of the Design Team as part of the background material they review prior to the actual visit.

The second stage of the visual reconnaissance tour usually occurs as a bus tour during the Friday afternoon of the weekend charrette, although teams have occasionally been driven around towns in all-terrain vehicles and even in a horse-drawn carriage during a rainstorm. The tour provides team members with an overall sense of the region under consideration as well as insights into more specific problems and potentials found at various locations. Team members typically seek background information and data from the tour leaders to deepen their understanding of the spatial environments. Team members will frequently return to specific sites to take additional notes and measurements, make preliminary sketches of buildings and significant architectural details, or just quietly absorb the special qualities of a given site while storing up images in anticipation of the Saturday charrette (see Figure 5.2).

The visual data generated by means of the visual assessment techniques and reconnaissance tours offer community designers another window through which to view and understand the community. Just as the image of the city is built up through layers over time, so, too, the image of a community is created through the addition of different layers of meaning and experience. One final layer, the meanings of community opinion, must be added to the slowly emerging picture of the community before community design can be effectively undertaken.

NOTES

1. Special thanks to Dale Helmich, formerly a Program Associate with the Main Street Program of the National Trust for Historic Preservation, for her insights into the forces affecting traditional Main Street districts.

2. The Minnesota Design Team wishes to thank Randy Hester for his participation in a Design Team visit and for sharing some of his important insights during his visit.

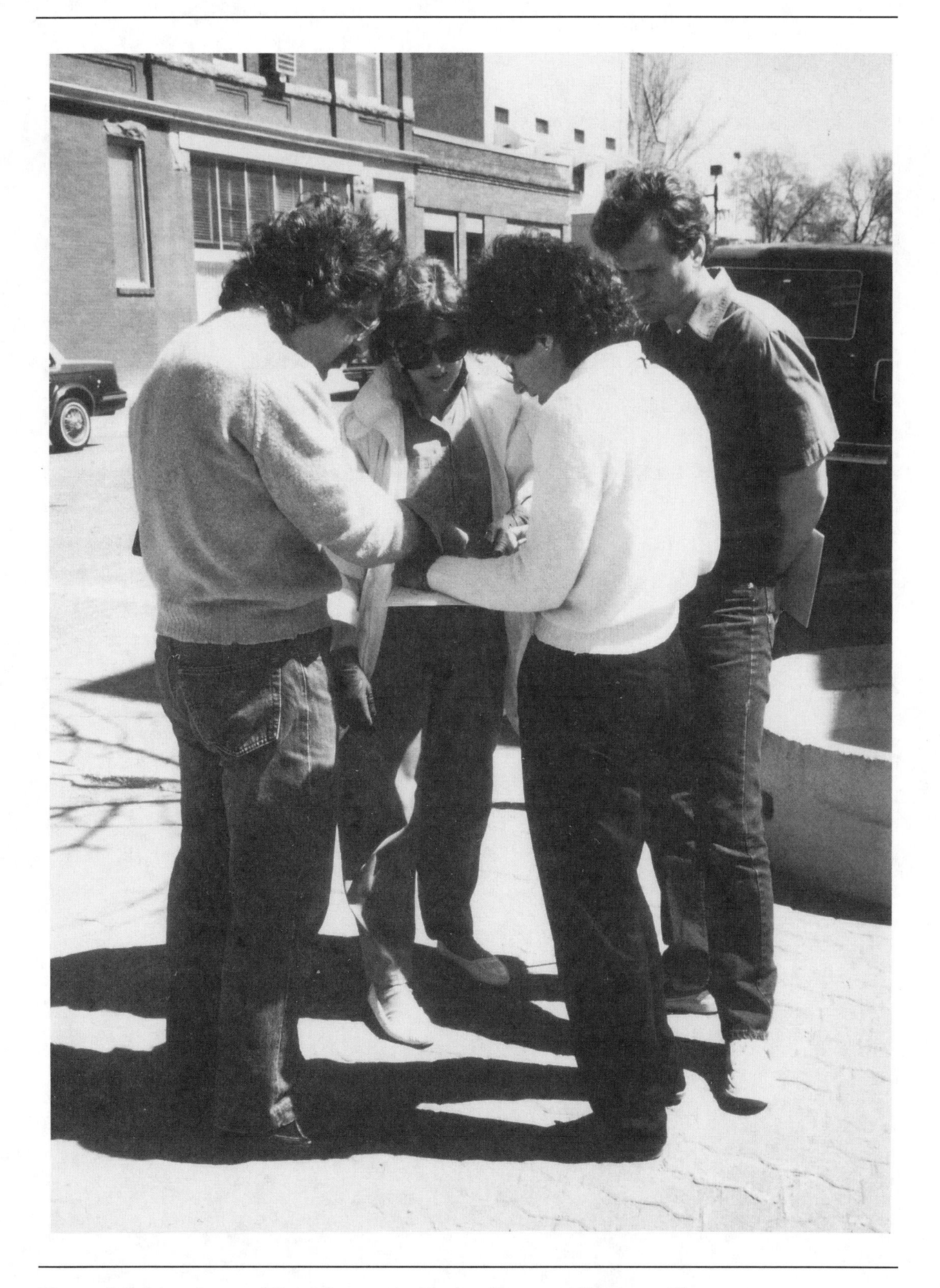

Figure 5.2 Members of the Minnesota Design Team at Work on Site

CHAPTER 6

GAUGING COMMUNITY OPINION

Adding the meanings held by different groups within a community to other research findings represents a necessary but still insufficient approach to community design. A community vision involves more than the sum of its competing special interests. The underlying question for community design still remains the most important one: Who actually speaks for a modern community in a post-industrial society, and can it find or create a common voice once again?

It can no longer be assumed that all or most citizens in contemporary American communities share the same system of meanings and values. As urban sociologist Louis Wirth (1938) noted in his now-famous essay "Urbanism as a Way of Life," modern Americans in metropolitan regions belong to many communities and play many, often contradictory roles in civic life. For example, the devoted father and active PTA member may work for a company that is planning to relocate its operations elsewhere, with possibly serious negative consequences for the local school district and municipality. With such divided loyalties pulling people in many competing directions, how, then, can the community designer and local decision makers discover what people in the community actually think about its current conditions and hope for its future?

USING SURVEY RESEARCH AS A COMMUNITY DECISION-MAKING TOOL

Many decision makers at state and national levels have come to depend heavily on opinion surveys to answer this question. However, survey research instruments such as questionnaires and public opinion polls, while useful tools, often measure only personal opinions instead of encouraging healthy and much-needed civic dialogue about fundamental community values. Evaluation researchers (Krueger, 1988) increasingly acknowledge the benefits of combining both qualitative and quantitative measures, resulting in stronger methodological approaches that strengthen the research design and its findings. Qualitative measures such as field research "will provide in-depth information into fewer cases," whereas quantitative procedures such as opinion polling "will allow for more breadth of information across a larger number of cases" (p. 38).

This distinction between qualitative and quantitative measures becomes especially important for purposes of making community design decisions. The more influence the research findings will exert on future planning and policy decisions, the greater the need for a variety of survey measures (Henerson, Morris, & Fitz-Gibbon, 1978, p. 15). Combining survey research with community visioning techniques offers valuable opportunities to gain a greater and more accurate representation of community opinion, overcome static positions, and create a deeper sense of community consensus in the process.

Does such a thing as a broad community consensus still exist in contemporary American society? What do members of a local community actually think? What is their attitude toward how their community is shaping up? The concept of a community attitude or opinion is, like many similar abstract concepts, a mental construct. It is "a tool that serves the human need to see order and consistency in what people say, think, and do, so that given certain behaviors, predictions can be made about future behaviors" (Henerson et al., 1978, p. 11). It is a tool that decision makers, faced with critical issues in their communities about future behaviors, need to sharpen and polish considerably for improved use in the community design process.

This tool of community opinion research is especially important for leaders of increasingly decentralized systems such as local communities. Otherwise, it becomes "virtually impossible for them to have any current knowledge of how policies are being administered or accepted" (Dunham & Smith, 1979, p. 37). Yet, local governments are currently very limited in their ability to gauge general community opinion on critical issues. Most decision makers have to fashion their tools for assessing general community opinion from a random assortment of personal contacts with individuals and groups, loud or persistent citizen complaints, newspaper editorials, and various letters to the editor (Webb & Hatry, 1973, p. 1). The best community leaders do it intuitively and quite well, but all of them would like to obtain a clearer picture of general community attitudes and values before adopting a strategy and making crucial decisions that will shape their communities for decades.

Such a picture of community opinion, however, can be easily distorted. As time becomes increasingly devoted to commuting and running household errands, civic involvement by the general public dwindles.[1] Most community leaders have likely experienced low turnouts at regular city council and planning commission meetings, even when important decisions are being discussed and voted on. This situation then allows powerful and vocal minorities to push their special agendas forward without significant public awareness or resistance. Opinion researcher Peter Graves believes that politicians really do want to listen to their constituents but find it difficult to know who to listen to with so many voices each pleading its own cause.[2]

Knowing who to listen to remains highly problematic for local decision makers. On the one hand, simply assuming that those who are silent agree with current trends and decisions overlooks the very real possibility that

> many people feel their views are not wanted, or would not make a difference if expressed, that some simply cannot or do not know how to make their views known, and that still others are reluctant, for personal or political reasons, to speak out. (Webb & Hatry, 1973, p. 8)

On the other hand, opinion polls designed to show "what the public really thinks" may not help decision makers very much either. Public opinion is often "the snapshot collection of . . . undigested views, our private 'takes' unshaped by any process of discussion and give-and-take with other perspectives outside our immediate lives" (Boyte, 1995, p. 418). In addition, attempts to measure citizen attitudes are often "blurred by peer group pressures, the desire to please, ambivalence, inconsistency. [and] lack of self-awareness" (Henerson et al., 1978, p. 13).

What is missing from the community design toolbox is a reliable method of not only gauging public opinion but of building a public consensus instead of gathering a mere collection of individual or special interests. Because citizens seldom know beforehand what they agree upon, they "usually only find [those agreements] through the process of ethical reflection" (Brown, 1990, p. 56).

Building a public consensus in modern communities characterized by increasing diversity of people, roles, and interests requires broad and deep citizen involvement to obtain the multiple perspectives needed for genuine ethical reflection. Because there is typically a lack of agreement about the nature of the community problem or what is needed for a successful solution, "heterogeneous group members must pool their judgments to invent or discover a satisfactory course of action" (Delbecq Van de Ven, & Gustafson, 1975, p. 5). Everyone is an expert in what the community means to them, and community designers desperately need their expertise. In Bennis's (1993) words, "democracy becomes a functional necessity whenever social systems are competing for survival under conditions of chronic change" (p. 22), such as the sea change of post-industrialization. The truth in terms of a community consensus is quite frequently a negotiated settlement.

BASIC APPROACHES TO ENGAGING THE COMMUNITY

Minnesota Design Team communities typically employ three basic approaches to the goal of bringing that citizen expertise into the community design process. Each approach possesses special advantages and weaknesses. Collectively, however, they reach into the inner workings of most communities and help create the critical mass of civic dialogue necessary to ensure the effectiveness of the community design process.

Interpersonal Communications

As in advertising, personal communication remains the most effective means. Word-of-mouth communications can take several forms. These include personal contacts by members of the community's steering committee, as well as presentations to community organizations such as Rotary clubs or school groups. For example, the City of Becker used a student intern to make presentations about the upcoming Design Team visit to all its major community organizations. Personal communications offer a speaker opportunities to gauge how an audience is reacting to the presentation, to ask and respond to questions immediately, and to generally exert the most impact in terms of creating strong commitment to the community design process.

Media Communications

Local media play an important role in generating community involvement in the design process. A number of communities have employed written communications with considerable effectiveness. For example, Lake Elmo used a series of regular mailings and newsletters to inform its citizens about the purpose and characteristics of the Design Team visit, while the local newspaper in Clearwater ran regular feature articles on the upcoming Design Team and examined some of the key community issues involved. Sophisticated desktop publishing software programs now make attractive and inexpensive publications much easier to prepare and to tailor to the needs of specific community audiences.

Increasing sophistication also characterizes how communities use media presentations to involve citizens in the community design process. For example, the Minnesota Design Team created a videotape titled *Postcards From Home* that enables viewers to gain an overall sense of its community design process. A number of communities have presented the videotape at meetings of local civic organizations to help generate widespread citizen interest. However, an increasing number of communities are now taking advantage of public access cable television to broadcast the Design Team video or even their own presentations about the community design process. For example, Little Falls videotaped the entire three-day design charrette and later televised selected parts over the local public access cable channel. As more communities develop their own Web pages and chat groups, additional opportunities will continue to emerge for reaching and involving new audiences.[3]

Survey Research

Once audiences are involved, however, community designers must find ways to capitalize on the involvement to get at those ever-elusive community attitudes. Community surveys provide an excellent starting point for getting a clear sense of general citizen attitudes (Fowler, 1993). According to noted social scientist William F. Whyte (1991a), surveys are especially useful "for the systematic measurement of attitudes, beliefs, and values" across a sample of the community (p. 269). Certain kinds of knowledge, such as generalizations and predictions about large groups of people, require the type of broad measurements best obtained through survey research (Backstrom & Hursh, 1963, p. 8). Research about citizen participation for the Urban Institute (Webb & Hatry, 1973, p. 1) showed that scientific surveys provide a unique means to test the public pulse and can be especially useful for allocating community resources and creating programs. "Survey findings should help determine budget priorities, identify needed changes in existing activities, and guide the physical location of facilities" (p. 15), all important aspects of the community design process. Community surveys can also serve as important communication mechanisms within our fragmented communities. Their anonymity can catalyze a fresh flow of information to decision makers beyond regular sources, and such surveys can also help decision makers discuss what they consider to be important issues with citizens (Dunham & Smith, 1979, pp. 54-55).

However, surveys have several serious limitations, which must be kept in mind by community designers. First, citizen surveys do not provide all the answers necessary for accurately gauging community opinion. They should also not be used "unless public officials and managers are willing and able to finance them adequately, formulate them carefully, and analyze their findings thoroughly" (Webb & Hatry, 1973, p. 66). Their effective use by decision makers "requires a real understanding of the kind of information surveys can provide" (Dunham & Smith, 1979, p. 61), namely broad community trends and citizen attitudes. Finally, according to Whyte (1991a), surveys are most effective when they are combined with qualitative methods such as observation and interviewing (p. 269).

Questionnaires

One of the most popular survey research methods is the questionnaire. A questionnaire is a list of questions created for obtaining information and opinion about a broad category of the community. The questionnaire is then mailed to potential respondents, who are asked to complete the questions and return them by mail; the rapid growth of e-mail and user groups could make future survey research questionnaires much easier and quicker to distribute and return than at present.

Questionnaires permit wide coverage of the community. By giving respondents a sense of privacy as they answer the questions, questionnaires also lessen the potentially biasing effects caused by the presence of interviewers and may increase the validity of citizen responses. However, response rates for questionnaires usually do not exceed 50%, raising serious concerns about their reliability. In addition, because little is known about those who do not respond, significant bias may still exist in the sample (Miller, 1991, p. 141) and leave important groups in the community without a voice in the process.

Nevertheless, community surveys can serve as highly effective tools for community design. Although surveys are often difficult to prepare, administer, and analyze, many Minnesota Design Team communities have effectively conducted their own surveys. One of the best examples of the effective use of community surveys comes from the American West. The Town Council of Breckenridge, Colorado, initiated a major communitywide survey (RRC Associates, 1993) to determine voter priorities. Breckenridge used both a mailback survey and a telephone survey to complement each other; it also measured opinions of community subgroups against the community as a whole. The findings clearly revealed that parking and affordable housing were prime concerns in Breckenridge, a highly attractive venue for winter skiing, and gave the town council some valuable insights for use in their future decision making.

Standardized Surveys

Another variation of the community survey is the use of a standardized survey form.[4] The Minnesota Design Team has developed its own community survey form (see Figure 6.1), based on its understanding of the four underlying community environments and its cumulative experience regarding the key characteristics of successful communities. Transformed into survey questions, these key characteristics provide an excellent general set of benchmarks for communities to use in assessing their current situation. Repeated over time at regular intervals as a longitudinal survey, it can also offer the community a clear agenda for future improvement or identify strengths it can build on. An ongoing survey program "can show whether citizens perceive progress or degradation in the quality of services and other trends" (Webb & Hatry, 1973, p. 17). In addition, a standardized survey form such as this one provides a community design organization like the Minnesota Design Team with a rich body of comparative data for both research and programming purposes. For example, survey results from previous Design Team communities have provided the impetus for several statewide conferences on important community design issues, such as heritage tourism and sustainable development.

Data obtained from citizen surveys can become highly valuable for community design purposes, but such data "needs to be analyzed along with other information rather than being considered self-sufficient for planning, policy, and program decisions" (Webb & Hatry, 1973, p. 65). Several weaknesses make citizen survey data insufficient by themselves for gauging community opinion. One weakness of opinion surveys is that they assume that individuals really know how they feel about these crucial subjects. They also assume that individuals form their opinions one at a time, in isolation from other issues (Krueger, 1988, p. 23). Opinion polls that ask people whether they agree or disagree with a particular action or policy consequently remove that issue from its broader context of decision making. No person is an island, nor is any decision facing a community. Communities and decision makers need better alternatives to current single-issue approaches. They need to measure priorities, not opinions, and to place public issues into a decision-making context involving trade-offs between policy choices. To govern is to decide.

Delphi Technique

Making such informed choices requires much more civic dialogue than opinion polling traditionally allows. However, most community

Instructions: Please rate your community in terms of the following indicators of community health. Circle your choice (1 being the lowest rating and 5 the highest) for each indicator.

A. The Natural Environment:

1 2 3 4 5	Awareness of its natural systems (e.g., waterways)
1 2 3 4 5	Emphasis on a high level of environmental quality
1 2 3 4 5	Organizations devoted to protecting the environment
1 2 3 4 5	Environmental education programs for young people

B. The Economic Environment:

1 2 3 4 5	Careful management of community tax dollars and resources
1 2 3 4 5	Attention to the physical infrastructure (e.g.,roads)
1 2 3 4 5	A vital downtown business district
1 2 3 4 5	A well-planned economic development strategy
1 2 3 4 5	An entrepreneurial economic development organization

C. The Social Environment:

1 2 3 4 5	Support for traditional institutions (e.g., schools)
1 2 3 4 5	Acceptance and support for racial and gender diversity
1 2 3 4 5	Support for developing new community leadership
1 2 3 4 5	Widespread community participation in decision making

D. The Cultural Environment:

1 2 3 4 5	Community care and pride (e.g., beautification efforts)
1 2 3 4 5	Evidence of community initiative and follow-through
1 2 3 4 5	Special public places and community traditions (e.g., festivals)
1 2 3 4 5	An overall attitude of community cooperation

Figure 6.1. Minnesota Design Team Healthy Community Indicators Survey

leaders dread the thought of bitter and lengthy public hearings that add little to community consensus and often create many hard feelings with long memories. Powerful or overbearing individuals can often dominate such sessions, whereas group pressures within a community can also keep many individuals not normally in the spotlight from speaking their thoughts freely on important community issues. Group and individual interests may also override the problem-solving process and prevent the desired consensus from emerging (Uhl, 1971, p. 8).

It is often difficult to hear the public in a public hearing. The Delphi Technique represents one alternative approach that is designed to help gauge community opinion and measure priorities more effectively. Named after the legendary oracle at Delphi used by the ancient Greeks to determine a critical course of action, Delphi Technique is a decision-making and prioritizing process developed by the RAND Corporation about 1950 to obtain a high degree of consensus on critical issues without face-to-face discussion and conflict. Delphi Technique seeks to determine community consensus without face-to-face meetings; it may be especially useful for dealing with bitterly divided communities. It involves a series of questionnaires mixed with controlled opinion feedback. The process can save time and money, and its anonymity allows respondents to think independently about tough issues and to consider their opinions within a broader decision-making context (Uhl, 1971, pp. 7-8).

In Delphi Technique, a staff team and decision makers typically develop the initial questionnaire based on their understanding of the important issues. Respondents may then be asked to evaluate the list on the basis of some criterion, such as ranking its priority for their purposes; they may also be asked to generate new issues or ideas of their own. They then return the completed questionnaire. The staff team summarizes the responses and develops a feedback report along with a second questionnaire. Respondents are then asked to vote independently on the priority of the revised ideas or rankings (Delbecq et al., 1975, p. 11), or else to explain their reasons for remaining in the minority view. Participants then receive an updated list, summary, minority opinion report, and a final chance to revise their opinions. This iterative process helps ensure serious reflection on the issues and typically generates a high degree of group consensus on important community issues.

Two versions of Delphi Technique help illustrate its comparative advantages for gauging community opinion. Several years ago, the Heartland Center for Leadership Development in Lincoln, Nebraska, undertook a Delphi study of the mayors for 150 towns in Nebraska with under 10,000 population (Kokes & Todd, 1990). The mayors were polled on the future of rural communities and asked to identify the changes they considered critical to long-term community survival.[5] This example clearly demonstrates the ability of Delphi Technique to allow consensus building to occur over long distances or when it is difficult to bring respondents together in a single setting. However, the larger the group of respondents, the more difficult and time-consuming the task of tabulating responses and writing survey reports becomes; this procedure may work best for engaging civic leaders who represent special interests rather than the entire geographic community in a process of civic dialogue.

A second variation of Delphi Technique helps to supplement the approach used by the Heartland Center. Called the Community Consensus Survey, it was developed by Graves Opinion Research to offer communities an alternative to traditional opinion polling. This procedure requires community respondents to weigh paired choices of local issues so that community priorities are clearly considered and revealed in the process. These paired choices deal with three critical elements of local decision making:

- The community vision of itself
- What should be done and who should do it
- How citizens would fund these measures

Typical of Delphi Technique, the Community Consensus Survey can involve several repetitions. It also allows considerable analysis of the opinions of various community subgroups, an important factor for local decision makers. Finally, decision makers receive a prioritized view of the issues showing intervals between issues in the rankings, with all issues compared to each other and prioritized according to responsibility and revenue support. Both examples of Delphi Technique provide local decision makers with intelligence about their community in a form that can be easily translated into policies and resource allocation.

The anonymity of survey responses that characterizes Delphi Technique represents both a strength and a weakness in terms of gauging community opinion. What survey research typically cannot include is a clear sense of the particular meanings that citizens of the community assign to their responses. Survey measures generally deal with many of the symptoms of community problems, whereas written comments and interview responses can help explain the effects themselves and sometimes even their causes (Dunham & Smith, 1979, p. 93).

Nominal Group Processes

Nominal group process can help community designers concentrate on the most pressing issues that the community has identified, thereby increasing the value of each survey technique. Nominal group process is another valuable approach to gauging community opinion and consensus building. It attempts to use group dynamics to its advantage rather than depending on anonymity for its success. Whereas groups have been shown to be superior to individuals in terms of generating ideas (Delbecq et al., 1975), it is also true that group interaction can inhibit the creative thinking needed for community problem solving and consensus building. Nominal group process attempts to combine the best of both anonymous responses and group processes.

Nominal group process involves a structured, large group meeting that begins with a period of idea generation. Members initially are asked to generate their individual ideas about key community issues; this activity ensures opportunities for everyone to participate. Each participant presents input, which is recorded in full view of the entire group.

Small groups are subsequently formed to critically examine the ideas and to work toward some kind of group consensus about them; independent voting on the ranking of these ideas may also occur (Delbecq et al., 1975. p. 8; Dunham & Smith, 1979, p. 122). For example, participants may be given three stick-on notes and asked to place them on their top three priorities; those ideas that garner the most adherents then move to the top of the community agenda.

Nominal group process works best as a follow-up to a more general survey. It concentrates attention on the most pressing community issues, probing beneath the surface and meanings of more generalized responses. This process also separates idea generation from the evaluation phase, minimizing the role of dominant individuals or groups and giving increased attention to each idea generated (Delbecq et al., 1975, p. 9).

Democratic Brainstorming as Nominal Group Process

The Minnesota Design Team employs its own version of nominal group process, which it more folksily calls democratic brainstorming. Democratic brainstorming enables the Design Team to focus on key issues that have gradually emerged through surveys and field research, building on and testing the community's system of meanings. A typical democratic brainstorming session will occur at a Friday evening town hall meeting (see Figure 6.2), typically preceded by a communitywide potluck dinner that provides food for thought as well as for the body. Participants are given color-coded pins or ribbons as they enter the meeting hall. Following the potluck dinner, they are directed by the team leaders to sit at tables identified with the color corresponding to their pin or ribbon. This procedure helps to shuffle the deck, so to speak, and create some new community groupings rather than the habitual ones.

The key to a successful democratic brainstorming session lies in the series of questions to which participants are asked to respond. Team leaders create these questions on the basis of the key issues that have emerged from the preliminary research and through Design Team observations of the community. For example, communities dealing with a deteriorated Main Street might be asked to describe three highlights of a renewed Main Street through the eyes of a future visitor to the town; communities that have problems with inclusiveness in decision making might be asked to list as many possible shareholders in the future of the community as they can think of. Questions can be tailored to the needs and issues facing each community; they may deal with the community's past, current situation, or desired future, but they all need to probe beneath the community's surface appearances to be truly successful.

Figure 6.2. A Typical Friday Evening Town Hall Meeting

Participants write their responses to each question anonymously on a blank notecard. After each person has completed responses on notecards, all the responses to each question are collected by Design Team members. The collected responses are then given to another group to read and respond to; no group evaluates its own responses. This additional level of anonymity eliminates defensiveness about one's ideas and encourages greater reflection by group participants.

People not only need to feel that their views are welcomed but are actually being heard by others. According to pioneering community designer Lawrence Halprin (Halprin & Burns, 1974), "a simple device for group listening is to record visibly what everyone says and, in feedback sessions, to allow each person the assurance that he [sic] is being listened to and his input is being valued" (p. 55). What Halprin refers to as group listening has several components in common with democratic brainstorming sessions.[6] One Design Team member acts as a recorder for the group, writing down each citizen response on a flip chart as it is read until all responses to each question are clearly recorded for all present to view. Another Design Team member acts as a discussion facilitator for the group. After all the responses have been recorded, the facilitator asks participants whether the responses agree with or differ from their own views.

The group facilitator encourages participants to explain their reactions to the responses as well as their own views regarding the question. Responding to the cards they have received from other community members instead of staunchly defending their own well-rehearsed positions typically allows group participants to explore new ideas and stretch their own thinking about key community issues. Group members seem to enjoy becoming amateur anthropologists, looking at community issues through the eyes of others and trying to decipher this fascinating new code on the card before them; the group process takes on the aspects of an intriguing game instead of a heated and disheartening public debate.

THE IMPORTANCE OF FEEDBACK FROM THE SURVEY RESEARCH PROCESS

Regardless of what form is used to gauge community opinion, a systems approach to community design requires that members of the general community quickly and clearly receive feedback about what has resulted from the process and their involvement. Business management specialist Warren Bennis (1993) points out that most methods of survey research simply collect information. However, the survey feedback approach actually intervenes in the community design process and changes the very character of the community system. "The survey-feedback approach is utilized," writes Bennis, "in order to gain this extra commitment via active participation in the research process" (p. 143). Susan Kendall Tillman (1995), director of the Chattanooga Venture community visioning process, reinforces Bennis's remark. According to Tillman, the citizen participation process "lifted our goals to a community agenda instead of one sector's agenda. It's amazing how the city has been able to make progress that way" (p. 5).[7]

Building on this insight into the need for feedback from citizen opinion, the results of democratic brainstorming are quickly returned by the Minnesota Design Team to the general community. Following the small group discussions, the group facilitators and recorders immediately post the responses of their group to each of the questions on the wall along with those of the other groups. This immediate and verifiable feedback encourages further comparisons among groups and individuals, rewards people for their participation, and stimulates additional discussions of issues that often continue far into the evening or even early morning hours.

Democratic brainstorming does not require a great deal of statistical analysis to be effective in gauging community opinion. Its purpose is not analytical precision but community consensus building. "For many evaluation reports, all you want is an anecdotal summary. . . . Your goal will be to detect the most frequently expressed opinions, and to include these in your report, directly quoting when possible" (Henerson et al., 1978, p. 170). Working with community volunteers, the Minnesota Design Team prepares a written summary report of the democratic brainstorming session. Results of the answers to the questions are categorized and tabulated, then printed out and made available to the general public prior to the Saturday evening presentation. Direct quotations are used as often as possible to help preserve their meanings; people, especially children, love to see their own words, which are often quite revealing of important community attitudes. Preston Design Team leader Michael Lamb reported that during democratic brainstorming, one person responded to the question, "In five years, what would you like Preston to be known for?" with the answer "Growth without change." His answer succinctly captured the strong desire of the community to take control of rapid change and bring it back into line with basic community values.

A much fuller picture of the community now begins to emerge after systematic attempts at gauging community opinion are combined with field research and participant observation. The Minnesota Design Team also gathers immediately after the Friday evening democratic brainstorming session to compare group responses and to search for emergent patterns and themes. These patterns and themes become crucial ingredients of the design work of the following day.

NOTES

1. This is a major theme of Robert Bellah et al. (Bellah, Madsen, Sullivan, Swidler, & Tipton, 1985), in *Habits of the Heart*. Bellah and his colleagues argue that "the pressure to keep moving upward in a career often forces the middle-class individual, however reluctantly, to break the bonds of commitment forged with a community" (p. 197). Their

work has been highly influential in public discourse on the concept of the changing American community.

2. I am grateful to Peter Graves for the thoughts and insights about community survey research and consensus-building he shared during our correspondence and phone conversations. His firm, Graves Research, conducts extensive community consensus-building projects in municipalities and is widely regarded as one of the leaders in this field.

3. The Minnesota Design Team recently inauguarated its own home page. To reach it, simply type `http://www.minnesotadesignteam.org`. The home page contains information about the Minnesota Design Team, as well as upcoming and recent visits. The goal is to encourage communication between the Design Team and communities, as well as eventually linking communities with one another to share ideas, problems, and resources. According to Roger Karraker (1993) in *MacWeek*, both the White House and the U.S. Congress now use e-mail to communicate with citizens and employ software programs to categorize citizen feedback.

4. Survey researchers typically emphasize the higher validity of the findings from random surveys. However, in many of the small towns served by the Minnesota Design Team, a strong effort is made to reach every person in the community with the survey and solicit their participation. The survey used this way becomes as much a means of community building as it is a means of survey research. See Fowler (1993) for a more in-depth treatment of applied social research and survey instruments.

5. See Chapter 3 for the specific rankings of the issues identified by the mayors and a discussion of their relationship to the global restructuring of the communities.

6. Like most design professionals, members of the Minnesota Design Team owe an enormous debt to the design ideas as well as the community involvement process created by Lawrence Halprin. His famous "motation" technique involving citizens in describing their movement systems is just one of the many design ideas borrowed by the Minnesota Design Team.

7. The City of Chattanooga reconvened the community visioning process after it discovered not only how successful it was in terms of community design projects but how it had altered the process of community decision making and its ability to involve diverse segments of the community.

CHAPTER 7

YOU'VE GOTTA HAVE CONNECTIONS

Community Design as a Healing Process

I conceive of no flourishing and heroic elements of Democracy in the United States, or of Democracy maintaining itself at all, without the Nature-element forming a main part—to be its health-element and beauty-element—to really underlie the whole politics, sanity, religion, and art of the New World.

Walt Whitman, *Specimen Days*

Communities are by nature complex cultural ecologies involving interactions of places with people and their values, institutions, and livelihoods. The powerful forces of post-industrialization have fundamentally altered these dynamic systems, forcing them to become in effect intentional communities against their will and to consciously reflect on their reasons and modes for existing.[1] Although the very complexity of our communities requires intensive, detailed study employing a wide variety of research techniques, it is equally true that communities grudgingly resist being reduced to isolated fragments such as their demographic characteristics or basic economic indicators.

Instead, communities must be understood and cared for as if they were indeed living organisms whose identities far exceed the sum of their individual parts. Expanding on this metaphor, community designer Christopher Alexander argues (Alexander et al., 1977) that the organic wholeness of a community can only be realized by successive acts of healing, understood in its root sense of making whole and reconnected. He writes that "every act of construction . . . must be made in such a way as to heal the [community]" (p. 22). A systematic approach using the wealth of background research created through the community design process can help reconstruct such meaningful connections between people and places, economic vitality and quality of life, heritage and future, and begin to heal our communities.

Community design as a healing process therefore involves fundamentally rethinking the nature of our communities. Just as human health involves integrating various elements such as body, mind, and spirit into a unified whole, healing communities requires a similar process of integration among different elements. Making connections from background research about what constitutes healthy community design to the situations of living communities represents an excellent starting point for this healing process.

MAKING CONNECTIONS TO THE CONCEPT OF BIOREGIONALISM

Although social research offers community designers many valuable tools and insights to help them understand the incredible diversity of how humans assign meaning to places, environmental research as well as many recent natural upheavals have clearly revealed how communities fundamentally depend on the natural world for their existence. Furthermore, the unique physical settings of communities both shape and limit their subsequent actions. Despite the tendency of post-industrial societies to ignore the constraints of natural environments upon human settlements and their activities, communities themselves have clearly identified many problems associated with such a narrow understanding. New connections have to be made between communities and their natural settings to create healthier future development patterns.

A valuable legacy of scholarship on community design elegantly addresses the problem of how to connect people and places more effectively than our current, unsatisfactory pattern. This scholarly tradition, called bioregionalism, represents the "road not taken" for community design in the twentieth century. Bioregionalism is an approach to community design based on deep awareness of and appreciation for the natural systems operating within a region as well as acknowledgment of human adaptations that have evolved over time from interactions with those natural systems. It offers valuable insights into the complexities of community systems as well as creative means for shaping healthier future development patterns.

The famous Scottish planner Sir Patrick Geddes established much of the foundation of the bioregional approach to city and regional planning. Geddes was initially trained as a biologist and possessed a deep sense of the organic relationships that exist in nature. He also conceived of cities as essentially living organisms that grew out of the surrounding countryside, or hinterland. His views that town planning could not be separated from rural planning led him logically to the concept of the city region. "Do we not see," wrote Geddes (1968), "and more clearly as we study it, the need of a thorough revision of our traditional ideas and boundaries of country and town?" (p. 28). Although post-industrialization has undermined his earlier analysis of city-hinterland economic relationships by freeing cities from economic dependence on their traditional hinterlands, Geddes's insights into their enduring ecological connections and the need for a regional vision of the metropolis have subsequently been ratified over time.

Geddes also contributed valuable insights to the concept of human economic activities within a community. He made a critical distinction between what he termed *vital economics,* a life-efficient environment capable of meeting basic human needs, and the *money economics* characteristic of a market economy. "We are hypnotized by money but have lost sight of economics," he wrote, "the real functioning of life, in real and energetic health, creating real and material wealth. Real wealth can only be created in a life-efficient environment" (p. 70).

American urbanist Lewis Mumford expanded upon Geddes's concept of bioregionalism in his own attempts to balance human and natural needs with the disruptive forces of modern technology such as the automobile. Although Mumford wrote widely and well on virtually all aspects of urban society, such as art, architecture, and literature, willingly violating what he considered the gentleman's agreement of academic specialists not to invade each other's fields, he was mainly interested in assembling the disparate pieces into meaningful patterns. His holistic approach was "akin to Geddes' ecological or synoptic view of life, which emphasizes the interplay among occupations, social organization, and physical environment" (Goist, 1972, p. 383).

Mumford unceasingly advocated what he termed a *biotechnic civilization,* in which regional planning would balance the demands of technology. It is very much a systems approach to understanding communities. "One model Mumford offers of . . . wholeness in life is the ecological balance sought by living organisms and the [ecosystems] which embody such efforts" (Goist, 1972, p. 380). Such a planning approach would take into account natural systems such as open spaces and human needs for settlement patterns on a human scale, very similar to the contemporary pleas of the New Urbanists to use the prototype of the village in designing new communities (Katz, 1994). Through a combination of visioning and detailed knowledge of the complexities of local regions, Mumford offered another path through bioregionalism that would help make communities whole once again.

The work of Scottish planner and landscape architect Ian McHarg has greatly advanced the working methods needed to effectively implement the bioregional vision put forward by Geddes and Mumford. McHarg's ideas about community design clearly reflect those of his intellectual predecessors in the field of bioregionalism. "If we can create the humane city," wrote McHarg (1969), in words that echo both Geddes and Mumford, "rather than the city of bondage to toil, then the choice of city or countryside will be between two excellences, each indispensable, each different, both complementary, both life-enhancing" (p. 2).

Although he was an eloquent and influential advocate for the theory of bioregionalism, McHarg's greatest contribution to the concept may involve his methodology for community design. This technique involves conducting and codifying extensive ecological inventories of the natural and cultural systems operating within and upon a community. These inventories make it possible to implement bioregionalism in a much more systematic fashion by suggesting connections between previously unrelated elements and predicting the impacts of development decisions on wider community systems.[2]

Concerns about the increasingly negative impacts of unrelated development decisions on communities have fueled growing interest in sustainability. A sustainable community represents a dynamic system in which all members and groups cooperate toward creating a shared vision that preserves the best of its heritage, optimizes present opportunities, and preserves vital resources for the use of future generations. This current emphasis upon sustainable communities represents the latest manifestation of the bioregional approach to community design.

Sustainability shares the holistic view of bioregionalists like Mumford toward the concept of a community. "It is the missing sense of ecology and the commons that makes places real, turns housing into dwelling, zones into neighborhoods, municipalities into communities, and ultimately, our natural environment into a home" (Van der Ryn & Calthorpe, 1986, p. xviii). It forces community designers to confront issues of human scale—for example, zoning that separates potentially compatible uses such as residential and retail and gives access only to automobiles—and to consider the unique qualities of a particular physical setting.

In particular, sustainability has added the device of community indicators to the tool kit of community designers. Community indicators are locally derived measures of human satisfaction, such as numbers of children participating in afterschool programs, visibility of the surrounding mountains or historic site, or the health of trout in local streams. Expanding on the traditional notion of economic indicators of community well-being, community indicators grow out of more subjective community values, Geddes's vital economics, and thus vary markedly among regions. Community indicators offer another valuable means, like McHarg's ecological inventories, to connect people more meaningfully to their natural settings and create the biotechnic civilization advocated by Lewis Mumford.

Bioregionalism has evolved considerably from its origins in the fertile imagination of Sir Patrick Geddes, yet it continues to offer a viable path toward community healing. The bioregional approach envisioned by these leading scholars provides the underlying framework for the recommendations of community designers such as the Minnesota Design Team, offering a "big picture" to place communities into a meaningful natural context.

MAKING CONNECTIONS TO COMMUNITY RESEARCH

Although the big picture offered by bioregionalism is a necessary foundation, making clear connections to the detailed research conducted about a specific community is equally important to successful community design. As gifted researchers like Geddes, Mumford, and McHarg clearly understood, seeing the big picture of a particular bioregion also requires detailed analysis of its unique characteristics. The specific data collected then become the living material for creating more meaningful connections between the communities and their natural settings. Three basic categories of data provide this essential material for the community design process.

Types of Maps

Maps ground the community design process in a detailed understanding of the characteristics of a particular place. Several basic types of maps are typically employed in a community visioning exercise such as a Minnesota Design Team charrette:

1. Most communities usually have *aerial maps* of their site that reveal broad patterns. Some of these patterns include natural systems such as hydrology and vegetation, whereas others demonstrate the effects of human activities such as farming on the land. For example, aerial maps often reveal the shift from the tight grid pattern characteristic of the original small town to the cul-de-sacs that characterize new, suburban subdivisions emerging on the edges of town. Such patterns often sug-

gest ways to reinforce community identity and connect it in ways that relate it more intelligently to its site.

2. *Topographical maps* provide another valuable tool to better understand the physical setting of a community. These maps reveal community land forms and patterns. Such information is especially useful for understanding the hydrology of a site and for identifying environmentally sensitive sites, such as those possessing steep slopes subject to erosion or low-lying areas more subject to flooding.

3. *Land use maps* provide useful background information about the development patterns of the community. Land use maps can either illustrate existing patterns of land use or where future development is projected to occur. Relating land use maps to aerial and topographical maps offers a useful checkpoint to address potential conflicts, such as development proposed for sensitive sites. Such comparisons can also suggest new opportunities for development that reinforces existing land uses, such as using waterfronts to attract people to Main Street districts.

4. *Zoning maps* embody the legal ordinances adopted by the community to implement its comprehensive plan. Zoning maps show what type of development is legally allowed in a particular area, or zone, of the municipality. Zoning maps should be compared to land use maps to determine if they effectively correspond to one another, or whether the developments allowed by the existing zoning actually work against what the community wants to achieve. For example, some communities would like to develop more lively Main Streets but are forbidden by zoning laws from allowing downtown residential development. Once again, comparison of zoning maps with other maps can reveal problems and potentials for the site.

5. Finally, *special district maps* provide more detailed information about the physical characteristics of unique community sites such as downtowns, waterfronts, historic districts, or park systems. These maps allow community designers to probe more deeply into the nature of these sites and to explore a variety of options for their futures.

Types of Quantitative Data

Not all data is quite as spatially oriented as maps. Several types of data provide quantitative measures of the community under study that help compare it to similar communities as well as broader system trends.

1. *Census information* provides a readily available source of data about community demographic characteristics. Because demographics represent a key factor in any community system, demographic changes need to be regularly monitored and considered in any community design process. Demographic data can reveal important information about the relative age of the community, its median income, or race and ethnicity. One useful technique using census data involves creating a population pyramid. The population pyramid shows the relative percentages of males and females in each demographic cohort (e.g., under 10 years of age). This graphic presentation can effectively suggest some emerging community concerns, such as child care and education or a need for housing for senior citizens.

2. Many states have assembled *community profiles* of official municipalities. Community profiles provide vital economic information, such as major employers, primary occupations, and developable land in the municipal boundaries. Because economic development has become such a pressing concern in almost every community undergoing a community design process, this type of information becomes extremely important background information. It suggests areas of community concerns as well as possible connections to their site that could create more sustainable economies. For example, the heavily wooded bluff country in southeastern Minnesota has used its topography and vegetation to create a major visitor attraction that has revitalized local economies through tourism and is now generating value-added businesses such as forestry and a crafts industry focused on woodworking.

3. *Survey results* offer additional quantitative measures of community well-being. Surveys can range from simple questionnaires to highly sophisticated polling instruments.[3] Some major types of surveys include

- simple random sampling;
- systematic samples;
- stratified samples; and
- area probability sampling.

The main value of survey research to the community design process involves its ability to provide a measure of statistical reliability to decision makers about public opinion on selected issues. It complements basic demographic and economic data and helps connect design recommendations more intelligently to existing social and political concerns in the community. For example, survey results that indicate a community lacks a good understanding of its natural setting could suggest forming an environmental education curriculum in the local school district or working with a consultant to create an environmental inventory.

Types of Qualitative Data

Qualitative data derived through a variety of different techniques help connect community designers to the more subjective values and meanings held by local citizens.

1. *Interviews* with citizens such as host families provide an opportunity for in-depth discussions around the breakfast table of what community members value as well as their predominant concerns. Interviews can be conducted by mail and telephone or door-to-door. Minnesota Design Team communities typically employ door-to-door interview techniques because of the relatively small populations of the communities under study.

2. *SWOT analysis* creates a clearer understanding of what a community values and fears by inviting representatives of organizations to identify the strengths and weaknesses they perceive in the community as well as what forces they see acting upon it.

3. *Focus groups* allow special populations such as children and the elderly who may not enjoy equal access to community decision making to discuss their unique situations and concerns at length in a more convenient setting than is normally provided for them.

4. *Visual analysis* helps community designers understand in three-dimensional terms physical features of the setting that community members especially value or dislike, adding considerably to their ability to make intelligent design recommendations.

5. *Nominal group process,* or democratic brainstorming as the Minnesota Design Team calls this discussion technique, gives virtually anyone and everyone in the town an opportunity to add their insights and issues to the general discussion about the community's desired future. By focusing on themes that have emerged from other data sources and by offering participants anonymity in their critiques, nominal group process allows participants to speak freely about key community issues and ensures the broadest possible spectrum of community opinion. The data provided from this town meeting format shed considerable light on the ethos of a particular community and are one of the key sources of background information for the ensuing design recommendations.

Triangulation

These three categories of data collectively embody the research technique of triangulation. Just as surveyers try to identify multiple reference points to orient themselves, community designers need to employ a variety of tools and measurements to obtain a more complete and accurate picture of a community than any single one could offer. Triangulation also allows community designers to connect the general concept of bioregionalism to the specific characteristics of particular communities, providing a wealth of possible combinations for creative designers to explore.

MAKING CONNECTIONS AMONG THE DISCIPLINES

The community participating in a visioning workshop is not the only one involved in creating new understandings of the meaning of community. The design professionals who work with a community are also making connections among themselves and to the data about the host community. The ancient axiom "Physician, heal thyself" applies equally to community design as a healing process. Just as community members explore the nature of their community in a number of ways, sharing insights and ideas from a variety of backgrounds and perspectives, community designers such as members of a Minnesota Design Team also undertake the demanding process of building their own community. Connecting their individual perspectives to the insights of team members from many other fields as well as to the unique aspects of the community being assisted both parallels and interacts with the overall process of community design.

Training in a design profession necessarily requires considerable specialization to achieve professional competency and certification. Minnesota Design Team members, however, are compelled by the time pressures of the design charrette format as well as by their shared concern for the well-being of the communities to work together and enlarge their conceptual frameworks of what constitutes a community. The pro bono publico mission of the Minnesota Design Team also promotes cooperation, because no individual benefits from a disconnected design concept. By working together as a team, design experts gradually enlarge their understanding of the "big picture," the interconnectedness of the unique design elements of the community, as well as their own appreciation of the complexity of communities in general.

Community Design Fields

A wide variety of fields and disciplines are represented in the Minnesota Design Team through its affiliate organizations, and each visit draws on this wealth of design expertise. The purpose of using a team approach is to increase the number of perspectives available to understand the dynamics of the community system being considered. However, the diversity of perspectives also requires negotiating a common design framework.

The architecture profession provides a necessary building block for community design. Architects require other team members to thoughtfully consider the character of existing and new construction, the special features of the built environment such as historic buildings and unique architectural details, and how these features can be integrated more effectively with one another. For example, an architect during the Becker Design Team visit illustrated how the community could surround one of its most historically significant older buildings with needed new construction in a manner that enhanced both old and new buildings.

Landscape architecture naturally belongs in a community design effort. Landscape architects bring special expertise about natural systems shaping the built environment, such as the types of soils or hydrology, as well as awareness of ways to relate people more functionally and enjoyably to their natural environment. For example, landscape architects and designers in Clearwater suggested an open-air farmer's market along the Mississippi River as a community focal point that would not be seriously threatened by occasional flooding.

City and regional planning is another key to successful community design. City and regional planners enable other team members to better understand proposed land uses as well as how existing plans and zoning regulations may inhibit or assist other aspects of the community vision for itself. For example, planners on the Lake Elmo Design Team pointed out several ways in which the community's existing zoning laws, by promoting the development of suburban-type subdivisions with their wide streets and large lots, worked against its clearly expressed desire to maintain a rural, small-town character and appearance.

Economic development has emerged as an increasingly important element in the community design process. As communities struggle to find productive new uses for land whose value has been altered by post-industrialization, economic developers help remind other team members to move beyond prohibiting certain uses of land to suggesting creative new ways in which communities can productively use their land. For example, members of the Clearwater Design Team suggested that the community combine its location on the confluence of two rivers with the presence of several woodworking shops and plastics manufacturers. Together these factors could help to create a recreation industry that focused on water recreation activities such as canoeing and kayaking and locally manufactured water craft.

Good community design also draws on many other specializations for expertise related to the unique issues in the particular community under consideration. These may include such fields as agriculture, horticulture, forestry, and tourism. As one of the fastest-growing industries in the world, tourism has become an increasingly important factor in community design. Many communities now look to tourism as the miracle cure for their loss of manufacturing activity, a clean and lucrative industry that brings visitor dollars into town. However, such an economic development strategy contains many hidden costs, so it becomes critically important to consider tourism in a "big picture" approach. Tourism specialists benefit from dialogue with other design professionals, and team members from those fields are forced in turn to rethink some of their assumptions. For example, a tourism expert on the Little Falls visit helped team members view the community as part of a regional tourist destination as well as a separate entity. This realization helped the other designers think about the impacts of visitors from outside Little Falls on historic buildings and the Main Street district, parks and trailways, and a host of related planning and economic development issues.

Creating a Common Vision

The previous example of thinking about tourism in a wider context illustrates how community design needs to continually struggle to move beyond narrow specialist frameworks. A series of feedback and discussion sessions before and throughout the visit help team members create this common understanding of the community, connecting the data in increasingly meaningful ways as they gradually elaborate the design framework through a process of dialogue and debate.

The first attempt at creating a common framework of understanding occurs at a pre-visit briefing session. This session, held about two weeks before the Design Team visit, brings team members together for the first time. Members typically introduce themselves and describe their special interests and expertise. Team leaders then provide background information such as census data, specific community concerns, and a slide presentation to illustrate some of the most obvious community design issues. This session becomes the first draft of a community design framework.

Following the SWOT analysis, focus groups, and a tour of the community, team members assemble for a Friday afternoon debriefing session. Members share impressions and insights from these activities as well as talks with their host families. Team leaders typically record these ideas on large sheets of newsprint so that team members can compare their understandings and identify emerging patterns and issues as well as points of disagreement. This session provides the second draft of the community design framework. Team leaders use the information and insights from this session to formulate several key questions used at the Friday evening town meeting that will, it is hoped, provide meaningful answers to the questions team members have about the community. For example, perceptions by team members that widespread concern exists about a declining Main Street often lead to including a question that might ask community members to describe three key features of Main Street to a potential visitor ten years from the present.

The Friday night town meeting, involving widespread citizen participation about their concerns and visions, is the heart of the community design process for the Minnesota Design Team. Consequently, it is invariably followed by a Friday night review session regardless of the lateness of the evening or the weariness of the team members. This session represents the third and perhaps most crucial draft of the community design framework, because it grows out of the team's heightened awareness of what the community is actually thinking. It summarizes the main themes that emerged from group discussions of each of the key questions, as well as how team members perceive the relative importance community members assign to these themes. For example, the Friday night town meeting in Clearwater demonstrated widespread interest in and support for revitalization of the underused but historic riverfront. Consequently, riverfront revitalization became the dominant theme of the subsequent design framework created during the Saturday design charrette.

The Saturday morning concept plan represents the fourth draft of the continually evolving community design framework. Team leaders, who have worked with the community for months by this time, usually put forward their suggestions about how to connect the various design elements to themes and concerns expressed during the Friday presentations and meetings. This framework of suggestions usually involves a storyboard, or series of related maps and pictures, that develops the themes in a systematic manner. Team members then offer their reactions to the suggestions, making needed additions, deletions, and revisions. For example, team members in Clearwater pointed out the need for including an easily read land-use map that would help people perceive relationships between the riverfront and the rest of the town. The Saturday morning concept plan is arguably the most difficult but most exciting aspect of the community design process. It attempts to connect the mass of collected materials, ideas, and comments into a coherent whole, providing a meaningful community context for design professionals to exercise their special knowledge and creativity.

MAKING CONNECTIONS AMONG THE DESIGN ELEMENTS

Systems analyst Russell Ackoff (1981) recommends that "once a mission is formulated, however tentatively, it is useful to specify the properties with which one would ideally like to endow the system being designed" (p. 110). Good community design helps communities specify the desired properties of their revitalized community system. Good community design goes far beyond creating attractive individual projects, as valuable as they may be to their communities. In the next stage of the community design process, design experts use a highly creative, intuitive approach to fashion an overall design framework. This new design framework synthesizes the extensive research and discussion that has occurred before and during the visit, attempting to connect all physical elements of the community under study into a new form that makes each part work more effectively and that physically expresses community values. If it effectively incorporates the value system of the community, this new design framework becomes a visible symbol of the community's mission, creating its desired future.

Key Community Design Elements

A good design framework understands a community as a system of systems and attempts to connect the community under study more effectively to its regional context. Because political boundaries often do not correspond very well to the ecological, social, or economic systems that make up a community, the first important connection in this healing process involves identifying and linking the community more consciously to the actual stakeholders in its future. Traditionally, this is one of the most difficult conceptual leaps for any community to make and even more difficult to implement.

Several examples illustrate the importance of including this connection in the community design framework. The City of Paynesville in central Minnesota had experienced considerable development outside its municipal boundaries in the surrounding townships, contributing to political divisions in the community. Through the community design process, identification of water quality concerns by township residents concerned about the lakes and city residents concerned about the Crow River led to a shared recognition that the community also consisted of the entire regional watershed that linked their futures.

The Minnesota Design Team proposed a Community Cooperation Council (CCC) to improve coordination of the many projects in the region. Over the next few years, the community became a model for community cooperation. Paynesville established joint powers agreements for a watershed district and police protection, created a regional hospital district, developed a shared hockey center, and built a community senior center that received a J. C. Penney Award for community volunteerism.

Once such regional shareholder relationships are recognized, the next step in the community design framework involves reconnecting the community to its own vernacular tradition. Far too often, local communities fail to appreciate the special characteristics they have evolved over time in response to their location and natural setting, allowing development to occur that does not take these factors in consideration. This type of placeless development often initially spurs the community to undertake a design process, so successful community design depends to a great extent on making the connection to its unique identity.

Two examples illustrate different ways in which community design can effectively make this connection to the vernacular tradition. The Lake Elmo design charrette clearly revealed that local citizens highly valued the farms and open spaces surrounding the original village. The design framework suggested by the team offered a number of alternative patterns that would allow future development to occur in the community while preserving those valued farms and open spaces. Lake Elmo eventually revised its planning and zoning regulations to encourage more concentrated development, which would in turn preserve the valued farmland and open space that gave the community so much of its identity. In Embarrass, the Design Team pointed out to this northeastern mining community the presence of a number of historic Finnish farmsteads created by the original homesteaders. Local residents, long accustomed to these objects in the landscape, gradually began to see them as a valuable part of their heritage and worthy of preservation. Sisu Heritage, Inc., the local heritage preservation organization, subsequently raised over $500,000 in grant monies to promote their preservation (see Figure 7.1).

Restoration of historic farmsteads convinced residents to build new community facilities in their vernacular pattern as well. A thriving arts and crafts industry, as well as considerable tourist interest, has now emerged in Embarrass. In 1987, the National Trust for Historic Preservation named Embarrass one of America's sixteen "Uncommon Places."

Another critical design relationship involves connecting the community more thoughtfully and effectively to its gateway entrances. First impressions are often the most lasting ones, and visitors often use these first impressions to build their image of the town. This important experience also affects permanent residents. Endless strips of fast-food franchises and gaudy advertising signs indicate that nothing special dwells here, that the town is simply another undifferentiated mass of thoughtless commercial development. A clear commitment to community design requires making a strong symbolic statement of community identity at its gateway.

Figure 7.1. Historic Finnish Farmsteads Were Preserved Near Embarrass

For example, the town of Hallock felt it needed to reinforce its physical presence in a flat, somewhat featureless environment in northern Minnesota. The Minnesota Design Team responded to this concern by recommending creating an inviting green gateway through a concentrated program of tree planting on the major roads leading into town. Hallock made tree planting a top priority for its 1990 Celebrate Hallock centennial celebration. Citizens conducted a river clean-up project, developed a walking trail system, and solicited funds to plant "heritage trees." In highly innovative fashion, Hallock negotiated with the state highway department to obtain trees that were being removed for a major transportation project elsewhere in the state to plant around a local trailer park. A community task force prepared a landscaping plan and worked with a local landscape contractor to implement this important community connection.

Another important means of connecting communities is through their systems of parks and open spaces. For many community designers, the park system represents the very backbone of comprehensive urban planning. Parks and open spaces form natural systems within the community that can link the dispersed fragments of contemporary development patterns into vital new relationships.

The City of Taylors Falls in north central Minnesota addressed and effectively overcame this problem of fragmentation. A heavily traveled state highway essentially bisected the town, cutting its central business district off from a major regional park with its thousands of annual visitors. The presence of heavy trucks on the highway also posed a major threat to pedestrians, especially children, attempting to cross. The Minnesota Design Team recommended creating a pedestrian underpass along the scenic St. Croix river to connect the park with the downtown district (see Figure 7.2).

The resulting connection between these key design elements removed a major safety hazard, brought visitors and new vitality to the business district, and added a significant amenity for the enjoyment of residents and visitors alike.

Connecting a community like Taylors Falls once again to its historic waterfront is especially vital to the community design process. Perhaps because water composes so much of our bodies or because of our evolution, people seem to need connections to water. Most communities originated at their locations due to the presence of bodies of water, so reaffirming this historic relationship also helps revitalize community identity.

For example, South Saint Paul was a historic industrial center on the Mississippi River, its economy based on the meatpacking industry. By the early 1980s, that industry had virtually collapsed, and the packing plants along the river had closed their doors, leaving South Saint Paul an economically depressed community. It became apparent that local citizens valued the river but had lost sight of its meaning for their lives. The Design Team proposed the formation of a new community organization called the River Environmental Action Program (REAP) to stimulate new awareness of the role of the riverfront in the life of the community. REAP successfully fought off a neighboring city that wanted to establish a landfill site on their border, then began the healing process of revitalizing the riverfront. REAP began annual clean-up efforts with hundreds of volunteers, negotiated a fishing pier with the Department of Natural Resources, developed a boat launch, and created a River Walkway on the site of a former sewage plant.

Like reestablishing its relationship to its waterfront, connecting the community to its historic Main Street is also vitally important to successful community design. These commercial districts were often the focus of retail and banking activities as well as a social center, where farmers and ranchers stocked up on supplies and conversation. However, the larger scale of auto-oriented development gradually replaced the pedestrian scale of the Main Street district. Downtowns of all sizes are now

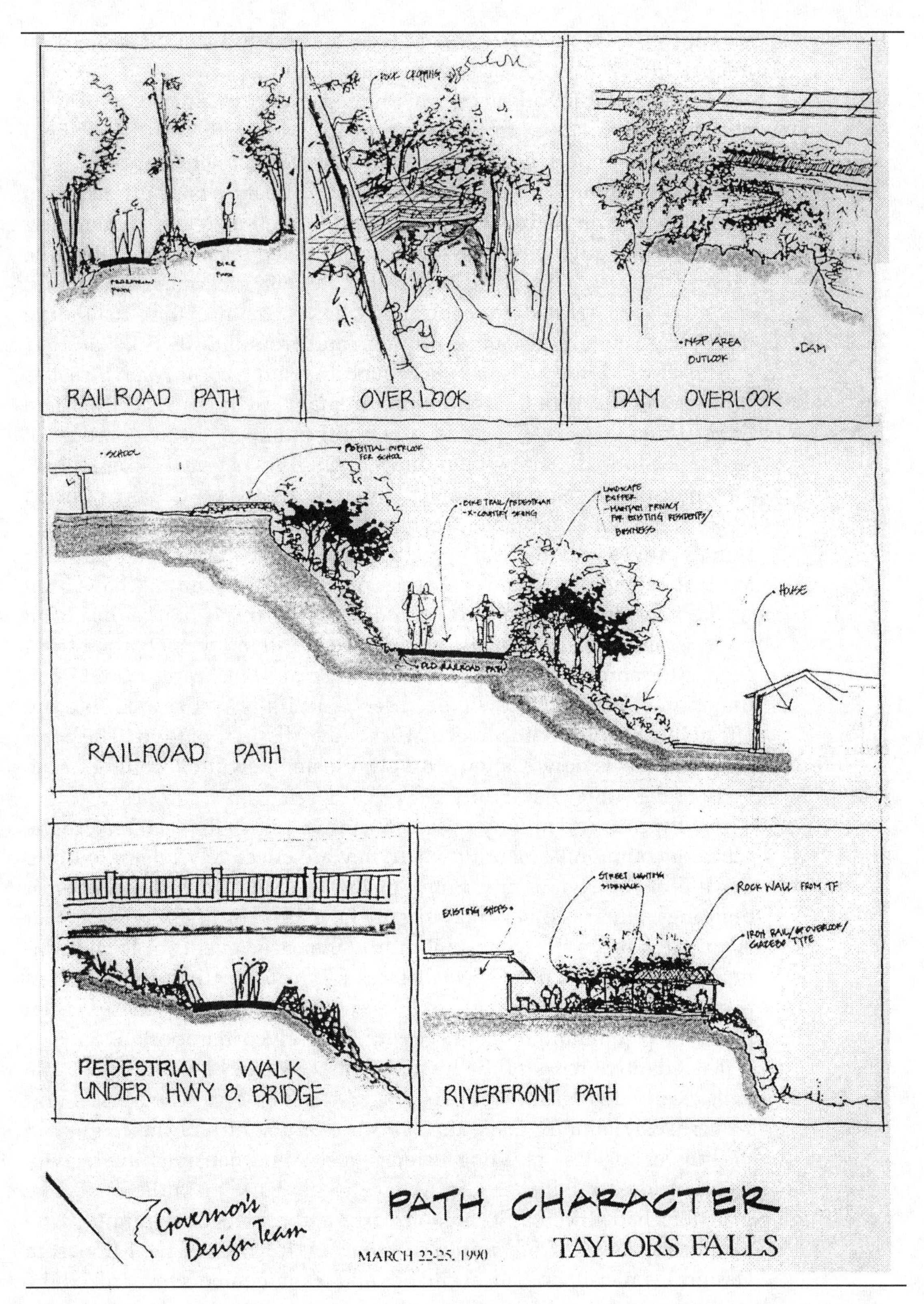

Figure 7.2. Sketches of a Proposed Pedestrian Underpass Along the St. Croix River in Taylors Falls

seeking ways to revitalize themselves, and many have discovered an important new function as a community meeting place.

Citizens of Little Falls in central Minnesota clearly indicated throughout the design charrette that their historic Main Street district held considerable importance to them, and downtown revitalization was an important community goal. Like most Main Streets, it had suffered from competition with highway development as well as regional shopping centers, and many of the aging buildings needed major renovation. The Minnesota Design Team suggested that the historic Main Street could serve as a focal point for the community, linking many of the outstanding natural and cultural resources found in and around Little Falls.

The City of Little Falls has subsequently reinforced its Main Street as the symbolic heart of the community in many exciting ways. The Main Street program was expanded and contributed to the restoration of several historic downtown buildings, including a highly successful senior housing development and the magnificent Cass Gilbert railroad depot. Little Falls implemented numerous landscaping and urban design improvements and is now in the process of acquiring easements for Main Street building facades. A series of large colorful murals have been painted on several Main Street buildings. A farmer's market has been revived, and thousands of visitors stream in to attend special promotions such as the annual Riverfest or the arts and crafts fair. Maple Island Park, the original site of the town, has been beautifully landscaped to more effectively link the Main Street district to the Mississippi River. The heart of Little Falls is now beating strongly, pulsing new life throughout the body of the entire community.

Finally, each community possesses special places that can restore the sense of community identity when they are effectively integrated into the life of the community. Some of these places are historic sites or buildings, whereas others may simply have been ordinary settings that acquired value to the community through special events or local traditions. In other cases, these special places are created by their communities to embody what they most value. Time-worn or brand-new, these special places help communities connect with themselves in important ways.

The adaptive reuse of the historic Musser-Weyerhauser site in Little Falls clearly demonstrated the power of special places. The Musser and Weyerhauser families made their fortunes at the turn of the century in the lumber business, building stately Queen Anne mansions on a heavily wooded ten-acre site next to Maple Island Park in Little Falls. The mansions had served as local landmarks and centers of community life for decades but had fallen into disrepair by the time of the Minnesota Design Team visit. Both local citizens and team members recognized the importance of this site to the community's past and future and made their preservation a top priority for the community design process.

The National Trust for Historic Preservation conducted an adaptive reuse study of the site and recommended that the mansions be used for continuing education and seminars. The City of Little Falls eventually obtained title to the site and mansions from family trusts and has now undertaken a major restoration effort of the properties. Building on the recommendations of the National Trust, the City of Little Falls is now hosting elderhostel programs at the site under the auspices of a local university. Participants in the program are helping to conduct historical research on the site as well as to catalog the contents; a lengthy waiting list has developed of participants eager to live and work at this unique site. At the same time, the City of Little Falls has attempted to make the site a major center for community life. Some of the activities at the site now include weddings and other special events, Christmas parties with sleigh rides, a Kiwanis Club fundraiser, and regular tours of the mansions and grounds.

A community is a complex living system that depends on mutual support from all its members for its continuing health. By showing the interrelationships of the many design elements, community design helps each of the elements function more effectively within the entire system and reinforces the community value structure as a whole.

MAKING CONNECTIONS TO THE COMMUNITY

The most sophisticated analysis and elegant design concepts are useless if a community does not use them in shaping its future. If the design recommendations are to provide useful guidance for future decision making, the community needs to discover its basic values incorporated into the new design framework. One key to successful community design involves connecting the proposed design framework to the prevailing values of the community by means of clear, powerful images that both reflect and renew the ethos of the community.

Anthropologist Clyde Kluckhohn (1949) once described anthropology as holding up a mirror to people to help them see their own culture in a new light. In much the same manner, community design involves holding up a mirror to communities to enable them to see themselves in new and, we hope, more meaningful ways. However, like individuals just emerging from sleep, communities do not always like what they see when gazing into a mirror. Good community design needs to both listen and lead, to show the community it can maintain its core values while adapting to new situations.

Two examples illustrate this important community design principle. South Saint Paul City Council member Lois Glewwe observed that the citizens of that community embraced the proposed design framework that grew out of that visit so eagerly because people could clearly see that "their ideas made it to those boards" displaying the proposed design solutions (Mehrhoff, 1995, p. 13). Another observer explained it more simply. Following the presentation of a proposal for special lighting on the bridge across the Mississippi River at Clearwater, a little girl in the audience shouted, "They listened to my idea!" To be successful, community designers need participants to clearly see themselves in the picture.

At the same time, a community design framework must help a community see possibilities for change as well as continuity. Psychologists and communications specialists alike have discovered the importance of a sense of structure in communicating complex concepts. One television commercial even proclaims that "Image is Everything." The systems approach to community design emphasizes considerable background research and community involvement in giving substance to a community design framework: Strong images are indeed crucial to helping community members both understand and remember the basic design principles and recommendations being put forward to help guide their community. Fetterman (1989) writes, "Unless the ethnographer couches the research findings in language the audience understands, the most enlightening findings will fall on deaf ears" (p. 22). Community designers should avoid superimposing inappropriate themes on communities;[4] design guidelines based on deeply held community values can become powerful images that are quickly adopted as the community's own and integrated into its decision making.

The Minnesota Design Team experienced the power that such a community-based design image can exert in the City of Lake Elmo. During the Friday morning SWOT analysis, a community representative talked about the absence of sewer and water lines in sections of the municipality. At first, some Design Team members assumed that local citizens regarded this as a negative condition; it gradually became apparent that many citizens did not want new sewer and water lines installed because of the development pressures they would create. Team members used this example to fundamentally rethink how Lake Elmo should approach development, focusing on "listening to the land" as the real message it had received from the community and as a clear organizing principle for future development (see Figure 7.3).

This powerful image helped Lake Elmo concentrate on protecting vital natural systems and open spaces, then directing development to "leftover" land that was not essential to the ongoing identity of the community. Lake Elmo has repeatedly drawn on this powerful image in wrestling with the difficulties of crafting new planning and zoning ordinances to implement its desired future.

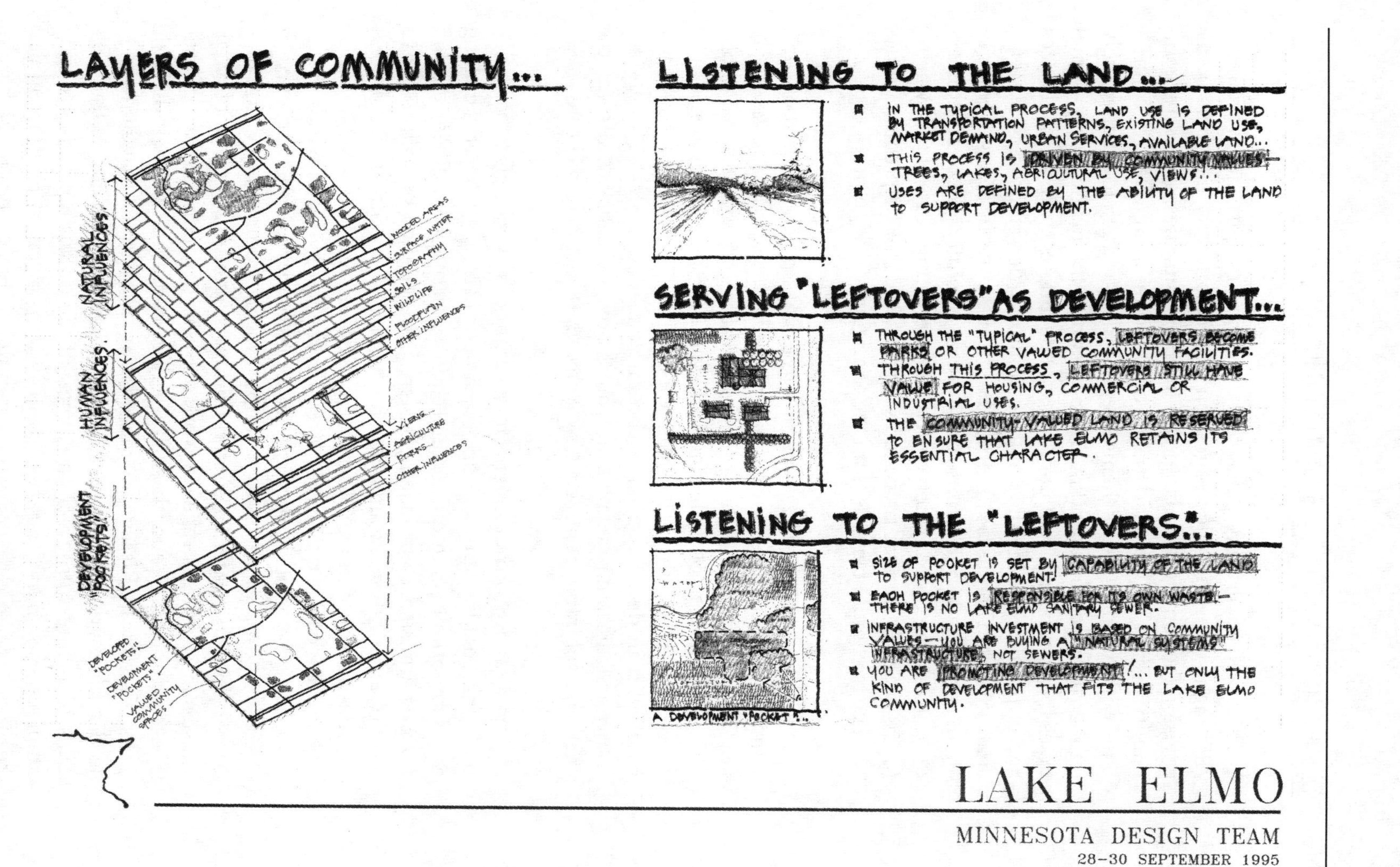

Figure 7.3. "Listening to the Land" in Lake Elmo

MAKING CONNECTIONS WITHIN THE COMMUNITY

Despite bold advertising claims to the contrary, image is not everything. Although strong graphic images can help community members better envision the common landscapes of their communities, successful community design ultimately depends on building more effective connections within communities themselves in order for them to act effectively on their new self-images (Fig.7.4). Sometimes the magic works; sometimes it doesn't. Each community responds differently to the challenges and opportunities identified in the community design process. Systems analyst Warren Bennis (1993) notes that "sometimes the changes brought about simply fade out because there are no carefully worked out procedures to ensure coordination with other interacting parts of the system" (p. 203). Some general procedures, or lessons, can help communities achieve maximum benefits from the community design process.

1. The first important lesson for successful implementation of community design involves *strengthening community networks and developing new leaders.* Because of the complexity of community systems, public officials alone cannot understand all aspects of a community nor meet all of its needs. Just as innovative businesses now involve all their employees in decision making and implementation strategies, the resources of the entire community must now be enlisted to meet the challenges of implementing a new community vision (see Figure 7.4).

A community's arrangements for a community visioning process like the Minnesota Design Team provide a useful starting point for making these important community connections. Steering committees created for the purpose of coordinating Design Team activities typically serve as a solid organizational foundation for helping to coordinate the process of implementation as well. Action research teams provide a valuable base of knowledgeable volunteers for conducting follow-up activities. At the Saturday evening design presentation, community members are encouraged to sign up to act on their support for the community vision they have helped to create. These sign-up sheets identify a large pool of enthusiastic volunteers eager to work on these new community projects. These volunteers frequently emerge as new community leaders. In Little Falls, Cathy van Risseghem moved from Design Team volunteer to leadership on downtown design improvements to director of the Little Falls Main Street program. In Paynesville, Tom Koshiol led a group of students known as the River Guard in a major effort to clean up and beautify the Crow River in that community. By expanding the base of volunteers, communities develop new reservoirs of skills and leaders.

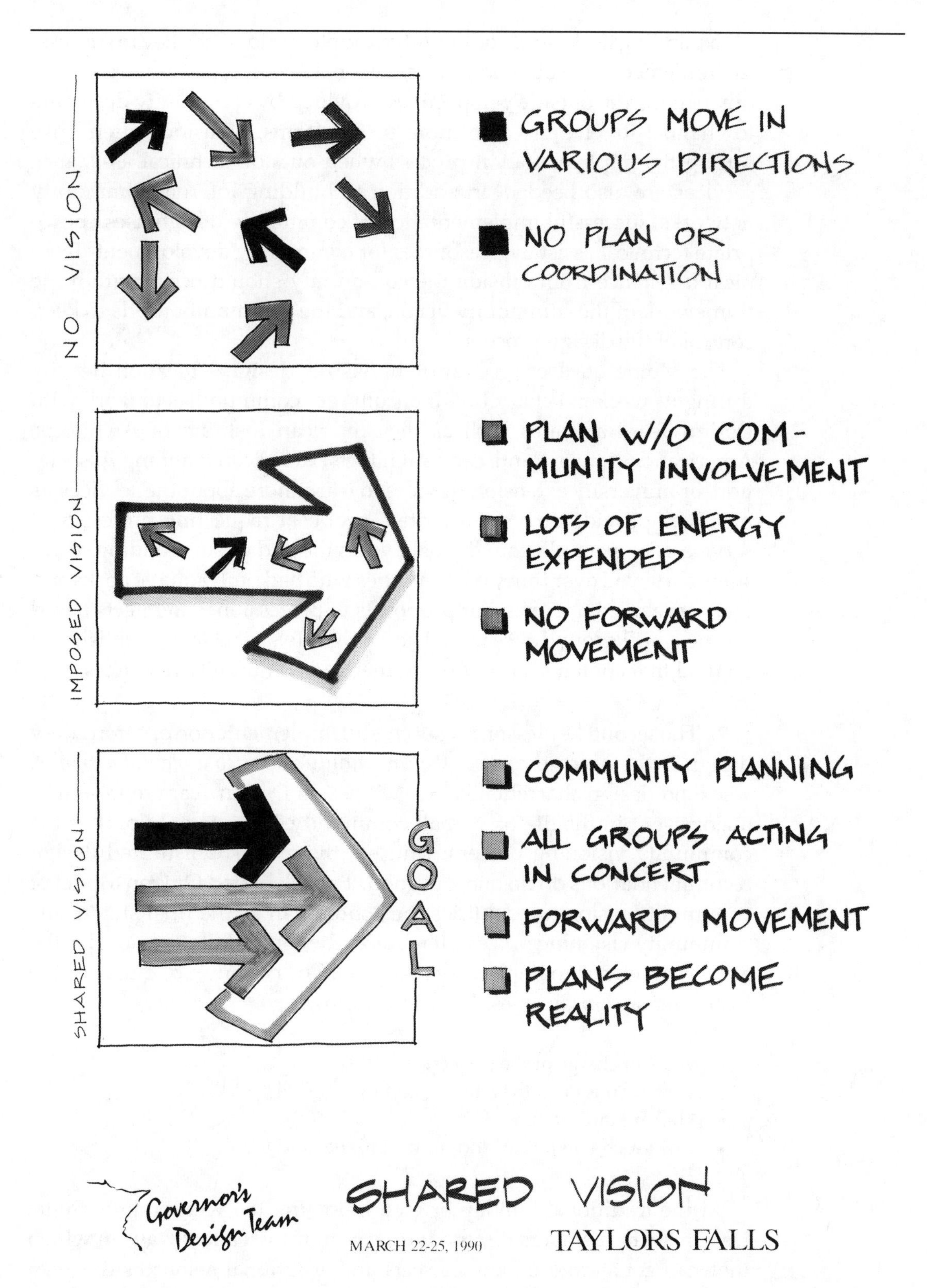

Figure 7.4. The Importance of Enlisting Everyone in Re-Visioning

Expanding the base of community volunteers holds the key to successful implementation of a community vision. However, just as the community needed the outside perspectives provided by community designers to channel its energies into more useful forms, occasions often arise during the implementation process when outside technical assistance services are also needed. In addition to building internal community networks, successful implementation of community design uses appropriate technical assistance resources for community development. Technical assistance from outside the community should occur within the framework of the community vision, and the community needs to keep control of the design process.

The Minnesota Design Team emphasizes design education heavily during its weekend charrettes. It encourages communities to work with Design Team affiliates such as the American Institute of Architects, American Society of Landscape Architects, American Planning Association, or university extension services to learn more about the services its members provide or how to write a proposal requesting professional services. For example, the City of Paynesville and its surrounding townships surveyed over thirty communities who had built aquatic parks and then developed a request for proposals to professional architects based on their background research. The Design Team has also published a manual that contains a list of recommended community resources.

2. The second key lesson for successful implementation of community design is the need to *connect the community vision to a plan of action.* A weekend design charrette like the Minnesota Design Team represents a major event in the life of a small community, but it is only part of the community visioning process. Unused, the most sophisticated design recommendations do no one any good. The celebrated Oregon Model of community visioning identifies the action plan as the final step of the community visioning process. It requires the community to prioritize the actions necessary to implement its vision. The community must be able to answer such basic questions as the following:

- Who's in charge of the project?
- What actions need to be taken, and by whom?
- What are the deadlines?
- How much will it cost, and how will it be paid for?

At the traditional Sunday brunch following the weekend charrette, Design Team members discuss with community leaders ways in which they can get started on key projects and additional resources they can call upon for help. Successful communities move quickly from vision to action. For example, the City of Lewiston engaged the services of

Minnesota Extension Services agent Roger Steinberg to help them prioritize the tasks required to implement the design recommendations. With this detailed framework in place, Lewiston was able to successfully coordinate public, private, and volunteer efforts and move forward on a wide variety of projects such as a new city park that grew out of the community vision.

3. The third key lesson for successful implementation of community design is to *connect citizens to the community vision with consistent and meaningful feedback.* As in many organizations, internal communications pose one of the greatest obstacles to effective action. Lack of the right information at the right time can prevent individuals and groups from participating in project activities, whereas poor communications can also weaken the sense of trust needed for joint action. Communities can enhance the implementation process through several simple forms of communications. A regular, easily accessible newsletter that focuses on implementation activities and upcoming opportunities is especially helpful for maintaining community support. Publicly displaying the results of the community visioning process, including citizen comments and design recommendations, is equally important. For example, some communities have laminated or dry-mounted design recommendations and displayed them in City Hall or other prominent community sites.

The Minnesota Design Team has developed a number of procedures to facilitate better communications within a host community. The community survey provides a valuable baseline of community indicators, including such factors as its skill at internal communications and its willingness to involve ordinary citizens in shaping a community vision. Next, the Design Team typically follows up the design charrette with a follow-up visit about six months later. As part of this follow-up visit, it asks the host community to complete an evaluation form about its subsequent actions (see Figure 7.5).

The answers to these questions provide useful information about the community's action plan. For example, the follow-up visit to Little Falls revealed that three separate community organizations had submitted grant proposals to the same funding agency. This revelation reinforced the need for more effective sharing of information and coordination of activities among the participants, and Little Falls subsequently became a model for other communities.

The Minnesota Design Team has also undertaken some large-scale evaluation procedures. Such macrolevel projects provided insights into the entire community design process, not just the experience of individual communities, and allowed overviews of the process of community design. The first of these macrolevel projects involved five-year retro-

1. How have you *communicated* with the community about projects growing out of the MDT visit?

2. Describe *volunteer efforts* in the community to implement projects that stem from the MDT visit:

3. Identify *new leaders* who have emerged in connection with MDT-related projects:

4. Explain how *community capacity* (e.g., new organizations) has expanded in response to the MDT visit:

5. Identify *outside resources* (e.g., political representatives, extension services, etc.) your community has utilized as a result of the MDT visit:

6. Describe any *fundraising* or *external grants* your community has successfully completed to carry out MDT-related projects:

Figure 7.5. Minnesota Design Team (MDT) Post-Visit Evaluation

spective studies (in 1987 and in 1992) of participating Minnesota Design Team communities. Participants were asked to evaluate the community design process and to suggest needed improvements. On both occasions, participants ranked technical assistance with implementation of the design recommendations as their top concern.

The second of these macrolevel projects involved statewide conferences that brought together representatives from Design Team communities around key themes. One of the conferences focused on heritage tourism, and the second addressed the theme of sustainable development. During both conferences, representatives from Design Team communities shared examples of their insights and successes with conference participants. This type of statewide or regional feedback system enables communities to build their own resource networks around topics of special interest.

CONNECTING THE CONNECTIONS

Management consultant Sally Helgesen (1995) has studied how successful organizations make network formation one of the keys to their success. She calls such organizations "webs," characterized by flexibility and constant adaptation to new situations. She writes,

> A web, though often configured to achieve a specific mission, plays a more important and lasting role. By emphasizing process as well as structure, by establishing new ways of approaching problems, of thinking, of connecting people, of giving them information and motivating them, a true web also helps to transform the organization of which it is a part. (p. 33)

The final lesson for successful community design, then, is to make the connection from simply doing projects to changing the entire pattern and processes of community decision making.

Two excellent examples illustrate this important community connection. A clear community vision can often streamline action planning and eliminate a great deal of wasted energy. Former Taylors Falls Mayor Steve Gall noted that "up until 1990 . . . [Taylors Falls] had no comprehensive plan for . . . development. The Minnesota Design Team was instrumental in getting that rolling" (quoted in Mehrhoff, 1995, p. 12). Since undertaking the community visioning process, Taylors Falls has formed a planning commission and created a new zoning ordinance. Taylors Falls worked successfully with the Minnesota Department of Transportation to build a pedestrian underpass along the St. Croix River, then completed a major new housing program for the city. Downtown pedestrian improvements, including historic streetlamps, have also been installed. As a result of community visioning, Taylors Falls operates in a totally different manner than before. According to Wade Vitalis, a local businessman who helped organize the Taylors Falls visit, "a real logical procedure has been established" that involves the entire community (quoted in Mehrhoff, 1995, p. 12).

South Saint Paul also illustrates how a community can make the connection to new ways of decision making. "The Minnesota Design Team taught us to think holistically," according to Darrol Bussler, South

Saint Paul's community education director, who helped arrange the visit. "It helped change the politics of the community and the way the community works" (p. 12). In 1990, the National Civic League chose South Saint Paul as an All-American City from 113 applicants. The league remarked that South Saint Paul's community-based development process "represented the best of creative, cooperative problem-solving." On August 6, 1990, President George Bush presented South Saint Paul with an All-America City Award. That hard-earned award just shows what a community can do when it has the right connections.

NOTES

1. The old river town of Valmeyer, Illinois, confronted this dilemma after being flooded by the Mississippi River back in 1993. Although its experience was more dramatic than those of many other small towns, the community design problems that Valmeyer had to address are a prototype for this general situation. See Watson (1996).

2. Geographic Information Systems (GIS) provide an excellent means for communities to take advantage of the wealth of community data assembled during the community design process. GIS are computer networks capable of holding vast amounts of data about geographic locations and applying the data to answer questions about their spatial relationships or the possible impacts of certain actions. To obtain the fullest benefit possible from the community design process, communities need to explore how to record and capture the wealth of new information about themselves and to apply it to future decision making. Local colleges and universities, as well as professional planning and landscape architecture firms, represent logical choices for helping communities take advantage of these new opportunities for achieving a bioregional understanding of themselves.

3. See Fowler (1993), especially Chapter 2 on sampling.

4. The misuses of the design process and the imposition of themes on communities are thoughtfully explored in a series of essays entitled *Variations on a Theme Park: The New American City and the End of Public Space* (Sorkin, 1992).

EPILOGUE

Genius Loci

In the summer of 1992, I received the gift of an opportunity to teach at the Centre for British Studies in Alnwick, England, located in Northumberland not too far from the border of Scotland. This wonderful experience allowed me to explore some of the most celebrated and beautiful landscapes in the world. These fabled English countryside estates created by Sir Humphrey Repton and Capability Brown back in the eighteenth century still provide value and delight as we approach the twenty-first century.

Such an approach to shaping communities and landscapes seems incomprehensible to many Americans, especially young people. Far too many students and young people in general have learned from us to focus only on short-term consumer wants; the concept of civic life has withered. Not surprisingly, many of them have also given up hope on the future, perceiving it as a fearful place of environmental disasters, a harsh and unforgiving economy, and widespread social disorder. The possibility of designing communities seems foreign to them.

A small group of students and professor living in a castle therefore provided the perfect vantage point to view American culture, in particular the design of our cities and towns. It helped to create a sense of community that is quite often missing in normal collegiate life and in American life in general. As a result of numerous field trips through cities and countryside, as well as helping the local community succeed with its beautification efforts for a Tidy Britain award, we began to talk about the values underlying these landscapes and their possible relevance to students' lives. Students noted with amazement that these experiences were beginning to change their value systems in significant ways.

We began to develop a checklist of what an enduring community would be like, and they suggested several important considerations that have guided me in my own efforts at building sustainable communities:

- Preservation of historic buildings and open spaces
- High level of care provided for civic spaces
- Finding new uses for historic sites (e.g., for film making)
- Emphasis on good work (e.g., arts and crafts)
- Respect for cultural diversity
- A long-term approach to resource management
- Education as an enduring source of meaning

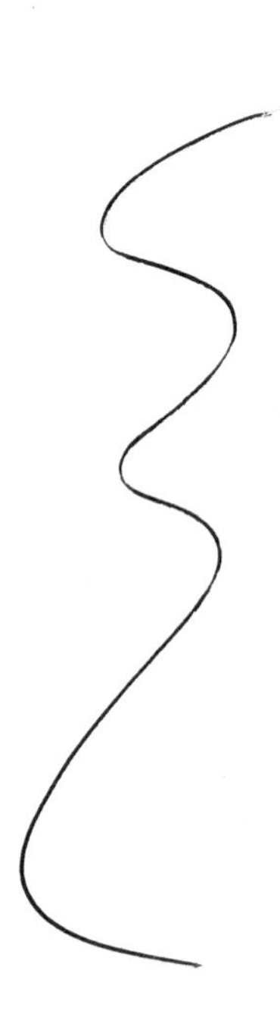

In the final analysis, then, I discovered that community design is not really about fashioning more handsome buildings, interesting views, or attractive landscapes. Community design is ultimately about empowering the citizens of local communities to shape their own preferred futures by acquiring and applying information and knowledge about their communities in a far more systematic, thoughtful, and democratic manner than current practice. Unlike many famous utopian designs for more enlightened physical and social landscapes, such as Frank Lloyd Wright's Broadacre City or the Ville Radieuse of Le Corbusier, community design does not depend for its ultimate success on the knowledge and control of one charismatic leader, like Wright's Master Architect. Instead, community design attempts the far more challenging task of building a new sense of what the ancient Romans termed *communitas,* or civic engagement, at the grassroots level.

Like the concept of communitas, community design ultimately aims at transmitting important, valued elements of our natural and cultural heritage to future generations for their own use and delight. I have been privileged to see many communities work at making their visions become realities, but it was a group of young children who truly taught me and other members of the Minnesota Design Team what community design is all about.

The River Guard is a group of children between the ages of eight and eleven who began to clean up the Crow River in the city of Paynesville shortly after the Design Team visit to their community (see Figure E.1). One of the key recommendations by the Design Team had been to attempt to recover the community's relationship to the river, from which the place had first sprung. The mayor of Paynesville, Joe Voss, asked local

Figure E.1. Members of the River Guard Cleaned Up the Crow River Frontage in Paynesville
Photograph courtesy of Paynesville Press; used with permission.

resident Tom Koshiol if he would consider helping to carry out that goal. Tom had not been heavily involved in community politics before the Design Team visit, but he enjoyed being outdoors and working with young people. He began organizing some of the youngsters who typically hung around the river to carry out this formidable task. Members of the River Guard first hauled out over two dozen truckloads of accumulated rubbish from along the banks of the river, then set to work making it a place of beauty both for themselves and for their entire community. They created a series of walking paths along the riverbank, then built an attractive footbridge across the river. A new spirit of place, what the Romans called *genius loci,* began to emerge where people had previously dumped their trash.

When the Minnesota Design Team returned to Paynesville over a year later for a follow-up visit, members of the River Guard proudly guided us around the newly transformed riverfront area. We followed them as they eagerly described their various projects and accomplishments, marveling at their sense of pride in reshaping their small piece of the community. After a while, we stopped at a rest area they had built along the trail, a very special place called Molly's Rest. Molly was the name of a beloved golden retriever who had belonged to one of the members of the River Guard. Molly had been buried in a quiet spot overlooking the river, and they had created and dedicated a lovely rest area of quiet dignity along the bank of the Crow River to her. The sense of peace and respect for the natural world at Molly's Rest was profound and absolutely overwhelming; in a very real sense, it had become a sacred place. I realized then that in its honest, simple way Molly's Rest represented the essence of what we were calling community design: creating meaningful connections between people and places that link the past and the future while helping to balance us against the forces of relentless change.

My thoughts slowly drifted back to when I was a little boy playing with my parents in that old park in north Saint Louis, how community design has given and continues to give meaning and order to my own life, and I gratefully recalled the classic words of another native of Saint Louis, T. S. Eliot:

> We shall not cease from exploration, and the end of all our exploration
> Will be to arrive where we started, and to know that place for the first time.[1]

NOTE

1. From T. S. Eliot, *The Complete Poems and Plays: 1909-1950,* New York: Harcourt & Brace, 1971, p. 145.

REFERENCES

Ackoff, R. L. (1967). Toward a system of system concepts. In J. Beishon (Ed.), *Systems behavior* (pp. 83-90). London: Harper & Row.

Ackoff, R. L. (1981). *Creating the corporate future: Plan or be planned for.* New York: John Wiley.

Agenda 21: The Earth Summit strategy to save our planet. (1993). Boulder, CO: Earthpress.

Alexander, C., Ishikawa, S., & Silverstein, M., with Jacobson, M., Fiksdahl-Kins, I., & Angel, S. (1977). *A pattern language: Towns, building, construction.* New York: Oxford University Press.

Ames, S. C. (1993). *A guide to community visioning: Hands-on information for local communities.* Salem, OR: American Planning Association.

Appleyard, D. (1979). The environment as a social symbol: Within a theory of environmental action and perception. *APA Journal, 45,* 143-153.

Appleyard, D., Lynch, K., & Myer, J. R. (1964). *The view from the road.* Cambridge, MA: MIT Press.

Arendt, R., with Brabec, E. A., Dodson, H. L., Reid, C., & Yaro, R. D. (1994). *Rural by design.* Chicago: Planners Press.

Argyris, C. (1985). *Action science.* San Francisco: Jossey-Bass.

Armstrong-Cummings, K., Barber, A., & Stutsman, R. (1996). Bringing it home: Sustainable practices in Kentucky. *Sustain, 1,* 8-11.

Backstrom, C. H., & Hursh, G. D. (1963). *Survey research.* Evanston, IL: Northwestern University Press.

Barker, J., Bruno, M., & Hildebrandt, H. (1981). *The small town designbook.* Jackson: Mississippi State University.

Barnett, J. (1996). *The fractured metropolis.* Boulder, CO: Westview.

Bartholomew, R. (1977). *Energy conservation in building design.* Monticello, IL: Council of Planning Librarians.

Bass, Rm E. (1993). *Mastering NEPA: A step-by-step approach.* Point Arena, CA: Solano.

Bateson, G. (1972). *Steps to an ecology of mind.* New York: Ballantine.

Beavis, M. A. (Ed.). (1990). *Ethical dimensions of sustainable development and urbanization: Seminar papers.* Winnipeg, Manitoba: Institute of Urban Studies, University of Winnipeg.

Bell, G. (1992). *The permaculture way: Practical steps to create a self-sustaining world.* London: Thorsons.

Bellah, R. N., Madsen, R., Sullivan, W. M., Swidler, A., & Tipton, S. M. (1985). *Habits of the heart: Individualism and commitment in American life.* Berkeley: University of California Press.

Bennis, W. (1993). *Beyond bureaucracy: Essays on the development and evolution of human organization.* San Francisco: Jossey-Bass.

Bensman, J. (1995). Community. *World Book Encyclopedia, 4,* 898.

Berry, W. (1993). *Sex, economy, freedom, and community.* New York: Pantheon.

Blakely, E. J. (1979). *Community development research: Concepts, issues, and strategies.* New York: Human Sciences Press.

Bogart, D. H. (1980). Feedback, feedforward and feedwithin: Strategic information in systems. *Behavioral Science, 25,* 237-249.

Bookchin, M. (1974). *Our synthetic environment.* New York: Harper & Row.

Botkin, J. W., Elmandjra, M., & Malitza, M. (1979). *No limits to learning: Bridging the human gap.* Oxford, UK: Pergamon.

Boulding, K. (1956, 1961). *The image.* Ann Arbor: University of Michigan Press.

Boyte, H. C. (1995). Public opinion as public judgment. In T. Glasser & C. Salmon (Eds.), *Public opinion and the communication of dissent.* New York: Guilford.

Bradley, N. L. (1996, June 23). A legacy filled with love of outdoors. *Minneapolis Star-Tribune,* pp. C19-C20.

Brewster, G. B. (1996). The ecology of development. *Urban Land, 55,* 21-25.

Brower, D., Godschalk, D. R., & Porter, D. R. (Eds.). (1989). *Understanding growth management: Critical issues and a research agenda.* Chapel Hill: The Urban Land Institute and the Center for Urban and Regional Studies at the University of North Carolina.

Brown, L. R. (1981). *Building a sustainable society.* Norton, NY: Worldwatch.

Brown, L., & Starke, L. (1994). *The state of the world, 1994. A World Watch Institute report on progress toward a sustainable society.* New York: Norton.

Brown, M. T. (1990). *Working ethics: Strategies for decision-making and organizational responsibility.* San Francisco: Jossey-Bass.

Buckley, W. (1967). *Sociology and modern systems theory.* Englewood Cliffs, NJ: Prentice Hall.

Buckley, W. (1968). *Modern systems research for the behavioral scientist.* Chicago: Aldine.

Camph, D. H. (1995). How sprawl costs us all. *Surface Transportation Policy Project Progress, 5,* 3-4.

Canter, D. (1995). Preface. In L. Groat (Ed.), *Giving places meaning* (pp. vii-viii). London: Academic Press.

Can we quote you on that? (1995). *On the Ground, 1,* 25.

Capra, F. (1975). *The tao of physics.* Berkeley, CA: Shambhala.

Castells, M. (1983). *High technology, space, and society.* Beverly Hills, CA: Sage.

Castells, M. (1989). *The informational city: Information technology, economic restructuring, and the urban-regional process.* Oxford, UK: Basil Blackwell.

Castells, M. (1996). Megacities and the end of urban civilization. *New Perspectives Quarterly, 13,* 12-15.

Chen, D. (1995). Sprawl: What you don't know can hurt you. *Surface Transportation Policy Project Progress, 5,* 5-6.

Chetkow-Yanoov, B. (1992). *Social work practice: A systems approach.* New York: Haworth.

Christenson, J. A., & Robinson, J., Jr. (Eds.). (1989). *Community development in perspective.* Ames: Iowa State University Press.

Clark, M., & Herington, J. (1988). *The role of environmental impact assessment in the planning process.* London: Mansell.

Clay, G. (1973). *Close-up: How to read the American city.* Chicago: University of Chicago Press.

Clay, G. (1980). *Close-up: How to read the American city* (2nd ed.). Chicago: University of Chicago Press.

Clay, G. (1987). *Right before your eyes: Penetrating the urban environment.* Chicago: Planners Press.

Clay, G. (1994). *Real places: An unconventional guide to America's generic landscape.* Chicago: University of Chicago Press.

Cox, H. (1966). *The secular city: Secularization and urbanization in theological perspective.* New York: Macmillan.

Cyberhood vs. Neighborhood. (1995). *Utne Reader, 68,* 52-75.

Daly, H., & Cobb, J. (1989, 1994). *For the common good: Redirecting the economy toward community, the environment, and a sustainable future.* Boston: Beacon.

Dead zone in Gulf traced to fertilizer use in state. (1996, June 17). *Saint Cloud Times,* p. 2A.

DeGrove, J. M. (1996). Managing growth in Minnesota. *Land Patterns, 1,* 1.

Delbecq, A. L., Van de Ven, A. H., & Gustafson, D. (1975). *Group techniques for program planning: A guide to nominal group and delphi processes.* Glenview, IL: Scott, Foresman.

Ditton, R. B. (1973). *National Environmental Policy Act of 1969 (P.L.91-190): Bibliography on impact assessment methods and legal considerations.* Monticello, IL: Council of Planning Librarians.

Downs, A. (1994). *New visions for metropolitan America.* Washington, DC: Brookings Institution.

Doyle, P. (1992, December 20). Rural folks relying more on city money. *Minneapolis Star-Tribune,* pp. 1A, 20-21A.

Drucker, P. (1986). The changed world economy. *Foreign Affairs, 64,* 774-791.

Drucker, P. (1993a). *The ecological vision.* New Brunswick, NJ: Transaction Publishers.

Drucker, P. (1993b). The new society of organizations. In *The learning imperative: Managing people for continuous innovation* (pp. 3-17). Boston: A Harvard Business Review Book.

Drucker, P. (1993c). *Post-capitalist society.* New York: Harper Business.

Dunham, R. B., & Smith, F. J. (1979). *Organizational surveys: An internal assessment of organizational health.* Glenview, IL: Scott, Foresman.

Dykeman, F. (Ed.) (1993). *Entrepreneurial and sustainable rural communities.*Proceedings of a conference held in St. Andrews-By-the-Sea, New Brunswick. Sackville, New Brunswick: Mount Allison University.

Ecologically informed designers: An interview with Ian McHarg. (1995). *On the Ground, 1,* 1-4.

Etzioni, A. (1968). *The active society.* New York: Free Press.

Featherman, D. L. (1992). Frontiers of social science and the SSRC. *Items: Social Science Research Council, 46,* 32-34.

Fetterman, D. M. (1989). *Ethnography: Step by step.* Newbury Park, CA: Sage.

Filipovitch, A. (1989). *Introduction to the city.* Dubuque, IA: Kendall-Hunt.

Fischer, C. S. (1977). *Networks and places: Social relations in the urban setting.* New York: Free Press.

Fischer, M. A. (1989). The practice of community development. In J. A. Christenson & J. W. Robinson, Jr. (Eds.), *Community development in perspective* (pp. 136-158). Ames: Iowa State University Press.

Fishman, R. (1977). *Urban utopias in the twentieth century.* New York: Basic Books.

Fishman, R. (1990). Megalopolis unbound. *The Wilson Quarterly, 14*(1), 25-45.

Flora, C. L., Spears, J. D., Swanson, L. E., with Lapping, M. B., & Weinberg, M. L. (1992). *Rural communities: Legacy & change.* Boulder, CO: Westview.

Forrester, J. W. (1970). *World dynamics.* Cambridge, MA: Wright-Allen.

Fowler, F. J. (1993). *Survey research methods.* Newbury Park, CA: Sage.

Frank, J. E. (1989). *The costs of alternative development patterns: A review of the literature.* Washington, DC: Urban Land Institute.

Fried, M. (1963). Grieving for a lost home. In L. Duhl (Ed.), *The urban condition* (pp. 151-171). New York: Clarion.

Friedmann, J. (1973). *Retracking America: A theory of transactive planning.* Garden City, NJ: Anchor.

Friedmann, J. (1993). Toward a non-Euclidean mode of planning. *APA Journal, 59,* 482-485.

Galbraith, J. R. (1995). *Designing organizations.* San Francisco: Jossey-Bass.

Gallagher, W. (1993). *The power of place: How our surroundings shape our thoughts, emotions, and actions.* New York: Poseidon.

Gans, H. (1962). *The urban villagers.* New York: Free Press of Glencoe.

Garreau, J. (1991). *Edge city: Life on the new frontier.* New York: Doubleday.

Geddes, P. (1968). *Cities in evolution.* New York: Howard Fertig.

Geddes, P. (1973). *City development: A report to the Carnegie Dunfermline Trust* (With an introduction by Peter Green). New Brunswick, NJ: Rutgers University Press.

Godschalk, D. R., & Brower, D. J. (1989). A coordinated growth management research strategy. In D. Brower, D. R. Godschalk, & D. R. Porter (Eds.), *Understanding growth management: Critical issues and a research agenda* (pp. 159-179). Chapel Hill: Urban Land Institute and Center for Urban and Regional Studies at the University of North Carolina.

Goist, P. D. (1972). Seeing things whole: A consideration of Lewis Mumford. *Journal of the American Institute of Planners, 38*(6), 379-391.

Gordon, D. (Ed.). (1990). *Green cities: Ecologically sound approaches to urban space.* Montreal: Black Rose Books.

Groat, L. (1995). Introduction: Place, aesthetic evaluation, and home. In L. Groat (Ed.), *Giving places meaning* (pp. 1-26). London: Academic Press.

American Planning Association, Oregon Chapter. (1993). *Guide to community visioning: Hands-on information for local communities.* (1993). Salem: Author.

Gunn, C. (1991). *Reclaiming capital: Democratic initiatives and community development.* Ithaca, NY: Cornell University Press.

Hall, P. (1984). *The world cities* (3rd ed.). London: Weidenfeld & Nicolson.

Halprin, L., & Burns, J. (1974). *Taking part: A workshop approach to collective creativity.* Cambridge: MIT Press.

Hammersley, M. (1992). *What's wrong with ethnography? Methodological explorations.* London: Routledge.

Hart, J. F. (1992). Population crisis in rural Minnesota. *CURA Reporter, 22,* 7-10.

Hawken, P. (1993). *The ecology of commerce: A declaration of sustainability.* New York: Harper Collins.

Hedman, R. (1984). *Fundamentals of urban design.* Washington, DC: Planners Press.

Helgesen, S. (1995). *The web of inclusion.* New York: Doubleday.

Henderson, H. (1988). *The politics of the solar age: Alternatives to economics.* Indianapolis, IN: Knowledge Systems, Inc.

Henerson, M. E., Morris, L. L., & Fitz-Gibbon, C. T. (1978). *How to measure attitudes.* Beverly Hills, CA: Sage.

Hester, R. T., Jr. (1985). Landstyles and lifescapes. *Landscape Architecture, 75,* 78-85.

Hester, R. T., Jr. (1989). Social values in open space design. *Places, 6*, 68-77.
Hester, R. T., Jr. (1995). Life, liberty, and the pursuit of sustainable happiness. *Places, 9*, 4-17.
Hough, M. (1995). *Cities and natural process.* London: Routledge.
Howard, R. (1993). Introduction. In *The learning imperative: Managing people for continuous innovation* (pp. xiii-xxvii). Boston: A Harvard Business Review Book.
Hunter, A. (1979). The urban neighborhood: Its analytical and social contexts. *Urban Affairs Quarterly, 14*, 267-288.
Hustedde, R., & Score, M. (1995). Force-field analysis. *Community Development Practice, 4*, 1-5.
Innes, J. E. (1992). Group processes and the social construction of growth management. *APA Journal, 58*, 440-453.
Jackson, J. B. (1984). *Discovering the vernacular landscape.* New Haven, CT: Yale University Press.
Jacobs, M. (1991). *The green economy: Environment and sustainable development and the politics of the future.* London: Pluto.
Jones, B. (1990). *Neighborhood planning: A guide for citizens and planners.* Chicago: Planners Press.
Karraker, R. (1993, June 21). D.C. embraces electronic democracy. *MacWeek, 7*(25), 1.
Kast, F. E., & Rosenzweig, J. E. (1970). *Organization and management.* New York: McGraw-Hill.
Katz, P. (1994). *The new urbanism.* New York: McGraw-Hill.
Keating, M. (1993). *The Earth Summit's agenda for change: A plain language version of Agenda 21 and the other Rio agreements.* Geneva: Centre for Our Common Future.
Keller, S. (1968). *The urban neighborhood.* New York: Random House.
Kemmis, D. (1993). The last best place: How hardships and limits build community. In S. Walker (Ed.), *Changing community* (pp. 277-287). Saint Paul, MN: Graywolf.
Keillor, G. (1990). *Lake Wobegon days.* New York: Viking Penguin.
Kluckhohn, C. (1949). *Mirror for man.* New York: Whittlesey House.
Kokes, K., & Todd, M. (1990). *Heartland Center study on the future of rural communities.* Lincoln, NE: Heartland Center for Leadership Development.
Kostof, S. (1987). *America by design.* New York: Oxford University Press.
Kotler, P., Haider, D., & Rein, I. (1993). There's no place like our place: The marketing of cities, regions, and nations. *The Futurist, 27*(6), 13-16.
Krueger, R. A. (1988). *Focus groups: A practical guide for applied research.* Newbury Park, CA: Sage.
Leo, J. (1993). Community and personal duty. In S. Walker (Ed.), *Changing community* (pp. 29-32). Saint Paul, MN: Graywolf.
Leopold, A. (1949). *A Sand County almanac and sketches here and there.* New York: Oxford University Press.
Lewin, K. (1951). *Field theory in social sciences.* New York: Harper.
Lewin, K. (1961). Quasi-stationary social equilibria and the problem of permanent change. In W. Bennis (Ed.), *The planning of change* (pp. 235-238). New York: Holt, Rinehart & Winston.
Lofland, J. (1971). *Analyzing social settings: A guide to qualitative observation.* Belmont, CA: Wadsworth.
Logan, J. R., & Molotch, H. (1987). *Urban fortunes: The political economy of place.* Berkeley: University of California Press.
Long, N. (1962, 1991). *The polity.* Chicago: Rand McNally.
Luke, T. (1993). Community and ecology. In S. Walker (Ed.), *Changing community* (pp. 207-221). Saint Paul, MN: Graywolf.

Luther, J. (1990). Participatory design vision and choice in small town planning. In F. Dykeman (Ed.), *Entrepreneurial and sustainable rural communities* (pp. 33-56). Proceedings of a conference held in St. Andrews-By-the-Sea, New Brunswick. Sackville, New Brunswick: Mount Allison University.

Lynch, K. (1960). *The image of the city.* Cambridge, MA: MIT Press.

Lynch, K. (1982). *What time is this place?* Cambridge, MA: MIT Press.

Majka, T. J., & Donnelly, P. (1988). Cohesiveness within a heterogeneous urban neighborhood. *Journal of Urban Affairs, 10,* 141-159.

Mandelker, D. R. (1984). *NEPA law and litigation: The National Environmental Policy Act.* Wilmette, IL: Callaghan.

Mannheim, K. (1945). *Man and society in an age of reconstruction.* London: Harcourt Brace.

Mantell, M., Harper, S. F., & Propst, L. (1990). *Creating successful communities: A guidebook to growth management strategies.* Washington, DC: Island Press.

McAllister, D. M. (Ed.). (1973). *Environment: A new focus for land-use planning.* Washington, DC: National Science Foundation.

McHarg, I. (1969). *Design with nature.* Garden City, NY: Doubleday.

Meadows, D. H., Meadows, D. L., Randers, J., & Behrens, W. W. (1972). *The limits to growth.* New York: Potomac Associates.

Mehrhoff, W. A. (1995). The Minnesota Design Team: A process of place-making. *Small Town, 26,* 4-13.

Melvin, P. M. (1985). Changing contexts: Neighborhood definition and urban organization. *American Quarterly, 37,* 357-367.

Miller, D. C. (1991). *Handbook of research design and social measurement.* Newbury Park, CA: Sage.

Miller, J. G. (1955). Toward a general theory for the behavioral sciences. *American Psychologist, 10,* 513-531.

Miller, J. G. (1978). *Living systems.* New York: McGraw-Hill.

Minnesota Environmental Quality Board. (1993). *A question of balance: Managing growth and the environment.* Saint Paul: Author.

Minnesota Planning Agency. (1993). *The view from Little Falls: Critical emerging issues.* Saint Paul: Author.

Moe, E. O. (1960). Consulting with a community system. *Journal of Social Issues, 15,* 29-35.

Morris, D. (1982). *The new city-states.* Washington, DC: Institute for Local Self-Reliance.

Mumford, L. (1968). *The urban prospect.* New York: Harcourt, Brace & World.

Myers, N. (1991). *The Gaia atlas of future worlds: Challenge and opportunity in an age of change.* New York: Doubleday.

Nasar, J. (1990). The evaluative image of the city. *APA Journal, 56,* 41-53.

Nelessen, T. (1995). Visions for a new American dream. *Surface Transportation Policy Project Progress, 5,* 3-6.

Norberg-Schulz, C. (1979, 1984). *Genius loci: Towards a phenomenology of architecture.* New York: Rizzoli.

Oldenburg, R. (1989). *The great good place.* New York: Paragon House.

Olsen, M. (1968). *The process of social organization.* New York: Holt, Rinehart & Winston.

Park, R., Burgess, E. W., & McKenzie, R. D. (1925). *The city.* Chicago: University of Chicago Press.

Park, R., Burgess, E. W., & McKenzie, R. D. (1967). *The city* (2nd ed.). Chicago: University of Chicago Press.

Perciasepe, R. (1996). The watershed approach. *Urban Land, 55,* 26-30.

Pierce, C. A. (1988). *Economics for a round earth.* New York: Vantage Press.

Power, T. M. (1996). *Lost landscapes and failed economies: The search for a value of place.* Washington, DC: Island Press.

President's Council on Sustainable Development. (1996). *Sustainable America: A new consensus for prosperity, opportunity, and a healthy environment for the future.* Washington, DC: Author.

Putnam, R. (1995). Bowling alone: America's declining social capital. *Journal of Democracy, 6*(1), 65.

Ramsay, M. (1996). The local community: Maker of culture and wealth. *Journal of Urban Affairs, 18,* 95-115.

Rapoport, A. (1969). *House form and culture.* Englewood Cliffs, NJ: Prentice Hall.

Rapoport, A. (1982). Identity and environment: A cross-cultural perspective. In J. S. Duncan (Ed.), *Housing and identity.* New York: Homes and Meier.

Real Estate Research Corporation. (1974). *The costs of sprawl: Environmental and economic costs of alternative residential development patterns at the urban fringe.* Washington, DC: Government Printing Office.

Redclift, M. (1987). *Sustainable development: Exploring the contradictions.* London: Methuen.

Redclift, M., & Benton, T. (Eds.). (1994). *Social theory and the global environment.* London: Routledge.

Relph, E. (1976). *Place and placelessness.* London: Pion.

Robertson, J. O. (1980). *American myth, American reality.* New York: Hill & Wang.

RRC Associates. (1993). *The Breckenridge community survey process.* Boulder, CO: Author.

Schatzman, L., & Strauss, A. L. (1973). *Field research: Strategies for a natural sociology.* Englewood Cliffs, NJ: Prentice Hall.

Schoemaker, P. J. H. (1995, Winter). Scenario planning: A tool for strategic thinking. *Sloan Management Review,* pp. 25-40.

Shore, E. (1993). The soul of the community. In S. Walker (Ed.), *Changing community* (pp. 3-8). Saint Paul, MN: Graywolf.

Sinha, A. (Ed.). (1995). *Landscape perception.* San Diego, CA: Academic Press.

Sorkin, M. (Ed.). (1992). *Variations on a theme park: The new American city and the end of public space.* New York: Hill & Wang.

Spradley, J. P. (Ed.). (1972). *The cultural experience: Ethnography in complex society.* Chicago: Science Research Associates.

Spirn, A. W. (1984). *The granite garden: Urban nature and human design.* New York: Basic Books.

Stedman, M. S., Jr. (1975). *Urban politics.* Cambridge, MA: Winthrop.

Stein, J. M. (Ed.). (1993). *Growth management: The planning challenge of the nineties.* Newbury Park, CA: Sage.

Sternlieb, G., & Hughes, J. W. (Eds.). (1975). *Post-industrial America: Metropolitan decline and inter-regional job shifts.* New Brunswick, NJ: Center for Urban Policy Research.

Sustainable Economic Development and Environmental Protection Task Force. (1995). *Common ground: Achieving sustainable communities in Minnesota* (A report to the Governor, the Minnesota Legislature, and the Minnesota Environmental Quality Board). Saint Paul: Minnesota Planning Office.

Swanstrom, T. (1996). Philosopher in the city. *Journal of Urban Affairs, 17,* 309-314.

Tillman, S. K. (1995, April). *Sustainability protects resources for future generations.* Washington, DC: National Renewable Energy Laboratory.

Toffler, A. (1980). *The third wave.* New York: William Morrow.

Trefil, J. (1994). *A scientist in the city.* New York: Doubleday.

Troughton, M. J. (1990). Decline to development: Towards a framework for sustainable rural development. In F. Dykeman (Ed.), *Entrepreneurial and sustainable rural communities* (pp. 23-30). Sackville, New Brunswick: Rural and Small Town Research and Studies Programme.

Tuan, Y.-F. (1977). *Space and place: The perspective of experience.* Minneapolis: University of Minnesota Press.

Uhl, N. P. (1971). *Identifying institutional goals: Encouraging convergence of opinion through the Delphi Technique.* Durham, NC: National Laboratory for Higher Education.

Van der Ryn, S., & Calthorpe, P. (1986). *Sustainable communities: A new design synthesis for cities, suburbs, and towns.* San Francisco: Sierra Club Books.

Vidich, A., & Bensman, J. (1968). *Small town in mass society.* Princeton, NJ: Princeton University Press.

Von Bertalanffy, L. (1968). *General systems theory.* New York: George Braziller.

Walker, M. (1994). Managers measure efficiency with citizen surveys. *American City & County, 109,* 47.

Walther, E. V. (1988). *Placeways: A theory of the human environment.* Chapel Hill: University of North Carolina Press.

Warren, R. L. (1963, 1972). *The community in America.* Chicago: Rand McNally.

Watson, B. (1996, June). A town makes history by rising to new heights. *Smithsonian Magazine,* 27(3), 110-120.

Webb, K., & Hatry, H. P. (1973). *Obtaining citizen feedback: The application of citizen surveys to local governments.* Washington, DC: Urban Institute.

Weiner, N. (1961). *Cybernetics* (2nd ed.). New York: John Wiley.

Wellman, B., & Leighton, B. (1979). Networks, neighborhoods, and communities: Approaches to the study of the community question. *Urban Affairs Quarterly, 14,* 363-390.

Whyte, W. F. (Ed.). (1991a). *Participatory action research.* Newbury Park, CA: Sage.

Whyte, W. F. (1991b). *Social theory for action: How individuals and organizations learn to change.* Newbury Park, CA: Sage.

Wiener, N. (1954). *The human use of human beings: Cybernetics and society.* Garden City, NY: Doubleday/Anchor.

Wirth, L. (1938). Urbanism as a way of life. *American Journal of Sociology, 44,* 1-24.

Wolf, F. A. (1981). *Taking the quantum leap.* San Francisco: Harper & Row.

World Commission on Environment and Development. (1987). *Our common future.* Oxford, UK: Oxford University Press.

Yanshen, W. (1995). Toward the twenty-first century. *Sustain, 1,* 5-7.

Zander, A. (1990). *Effective social action by community groups.* San Francisco: Jossey-Bass.

INDEX

ABOUT THE AUTHOR

W. **Arthur Mehrhoff** teaches American studies and local and urban affairs in the Center for Community Studies at Saint Cloud State University in Saint Cloud, Minnesota. A native of Saint Louis, Missouri, he earned a master's degree in urban affairs at Washington University in Saint Louis and a doctorate in American studies from Saint Louis University. His M.A. thesis looked at the role of citizen participation in environmental impact assessment, a theme continued in this work.

He worked in community organizing, community economic development, and urban design and was a city planner for the City of Saint Louis Community Development Agency. He also has worked as a freelance writer and has published scholarly articles in a variety of journals such as *Urban Resources, Urban Affairs Quarterly, Humanities Education, Minnesota Cities, The Canadian Review of American Studies,* and *Environment and Planning.* His first book, *The Gateway Arch: Fact & Symbol,* examined the evolution of American cultural attitudes toward the natural environment from our origins to post-modern America.

Since moving to Minnesota, Dr. Mehrhoff has taught numerous urban studies courses such as environmental design, housing and neighborhoods, and community development. In addition to teaching, he has been actively involved in community development activities throughout the state. He has led five Minnesota Design Teams and has served for many years as the community relations coordinator for the Minnesota Design Team Steering Committee.